MATHEMATICAL MODELING
of Earth's Dynamical Systems

MATHEMATICAL MODELING
of Earth's Dynamical Systems

A Primer

Rudy Slingerland
and Lee Kump

PRINCETON UNIVERSITY PRESS · PRINCETON AND OXFORD

Jacket illustration by Scott R. Miller, The Pennsylvania
State University. *Comparison of results from a landscape
evolution model (CHILD) simulating the Siwalik Hills
of India and Pakistan with an oblique aerial photo of the
same region. The view is toward the west, and the big
river is the Karnali. There is no vertical exaggeration.* For
details see Scott R. Miller and Rudy L. Slingerland (2006),
Topographic advection on fault-bend folds: Inheritance of
valley positions and the formation of wind gaps, *Geology*,
34(9): 769–772, dpo: 10.1130/G22658.1.

Library of Congress Cataloging-in-Publication Data

Slingerland, Rudy.
    Mathematical modeling of earth's dynamical systems : a
primer / Rudy Slingerland and Lee Kump.
        p. cm.
    Includes bibliographical references and index.
    ISBN 978-0-691-14513-6 (hardcover : alk. paper) —
ISBN 978-0-691-14514-3 (pbk. : alk. paper) 1. Gaia
hypothesis—Mathematical models. I. Kump, Lee R. II. Title.
    QH331.S55 2011
    550.1'5118—dc22

                                        2010041656

British Library Cataloging-in-Publication Data is available

This book has been composed in Sabon
Printed on acid-free paper. ∞
Printed in the United States of America
10 9 8 7 6 5 4 3 2 1

Contents

Preface   xi

1

Modeling and Mathematical Concepts   1
   Pros and Cons of Dynamical Models   2
   An Important Modeling Assumption   4
   Some Examples   4
      Example I: Simulation of Chicxulub Impact
         and Its Consequences   5
      Example II: Storm Surge of Hurricane Ivan
         in Escambia Bay   7
   Steps in Model Building   8
      Basic Definitions and Concepts   11
      Nondimensionalization   13
      A Brief Mathematical Review   14
   Summary   22

2

Basics of Numerical Solutions by Finite Difference   23
   First Some Matrix Algebra   23
      Solution of Linear Systems of Algebraic
         Equations   25
   General Finite Difference Approach   26
      Discretization   27
      Obtaining Difference Operators by
         Taylor Series   28

Explicit Schemes   29
Implicit Schemes   30
How Good Is My Finite Difference Scheme?   33
Stability Is Not Accuracy   35
Summary   37
Modeling Exercises   38

3

Box Modeling: Unsteady, Uniform Conservation of Mass   39

Translations   40
Example I: Radiocarbon Content of the
Biosphere as a One-Box Model   40
Example II: The Carbon Cycle as a
Multibox Model   48
Example III: One-Dimensional Energy
Balance Climate Model   53
Finite Difference Solutions of Box Models   57
The Forward Euler Method   57
Predictor–Corrector Methods   59
Stiff Systems   60
Example IV: Rothman Ocean   61
Backward Euler Method   65
Model Enhancements   69
Summary   71
Modeling Exercises   71

4

One-Dimensional Diffusion Problems   74

Translations   75
Example I: Dissolved Species in a
Homogeneous Aquifer   75
Example II: Evolution of a Sandy Coastline   80
Example III: Diffusion of Momentum   83
Finite Difference Solutions to 1-D Diffusion
Problems   86
Summary   86
Modeling Exercises   87

| 5 |

Multidimensional Diffusion Problems 89

Translations 90

Example I: Landscape Evolution as a
2-D Diffusion Problem 90

Example II: Pollutant Transport in a
Confined Aquifer 96

Example III: Thermal Considerations in
Radioactive Waste Disposal 99

Finite Difference Solutions to Parabolic
PDEs and Elliptic Boundary Value
Problems 101

An Explicit Scheme 102

Implicit Schemes 103

Case of Variable Coefficients 107

Summary 108

Modeling Exercises 109

| 6 |

Advection-Dominated Problems 111

Translations 112

Example I: A Dissolved Species in a River 112

Example II: Lahars Flowing along Simple
Channels 116

Finite Difference Solution Schemes to the
Linear Advection Equation 122

Summary 126

Modeling Exercises 128

| 7 |

Advection and Diffusion (Transport) Problems 130

Translations 131

Example I: A Generic 1-D Case 131

Example II: Transport of Suspended
Sediment in a Stream 134

Example III: Sedimentary Diagenesis:
Influence of Burrows 138

Finite Difference Solutions to the Transport
Equation   143
QUICK Scheme   144
QUICKEST Scheme   146
Summary   147
Modeling Exercises   147

8

Transport Problems with a Twist:
The Transport of Momentum   151
Translations   152
Example I: One-Dimensional Transport
of Momentum in a Newtonian Fluid
(Burgers' Equation)   152
An Analytic Solution to Burgers' Equation   157
Finite Difference Scheme for Burgers'
Equation   158
Solution Scheme Accuracy   160
Diffusive Momentum Transport in
Turbulent Flows   163
Adding Sources and Sinks of Momentum:
The General Law of Motion   165
Summary   166
Modeling Exercises   167

9

Systems of One-Dimensional Nonlinear Partial
Differential Equations   169
Translations   169
Example I: Gradually Varied Flow in an
Open Channel   169
Finite Difference Solution Schemes for
Equation Sets   175
Explicit FTCS Scheme on a Staggered Mesh   175
Four-Point Implicit Scheme   177
The Dam-Break Problem: An Example   180
Summary   183
Modeling Exercises   185

10

Two-Dimensional Nonlinear Hyperbolic Systems   187
    Translations   188
        Example I: The Circulation of Lakes,
            Estuaries, and the Coastal Ocean   188
    An Explicit Solution Scheme for 2-D
        Vertically Integrated Geophysical Flows   197
        Lake Ontario Wind-Driven Circulation:
            An Example   202
    Summary   203
    Modeling Exercises   206

Closing Remarks   209

References   211

Index   217

Preface

This book is a modeling primer, or first book of instruction, for geoscientists. Our objective is to teach graduate and advanced undergraduate students the skills necessary to represent complex Earth systems with mathematical and computational models that provide enhanced insight into processes and their products. It is written for students developing an expertise in the earth sciences and not for experienced environmental modelers and applied mathematicians. We assume only that the reader is familiar with the principles of physics, chemistry, and geology and has had a year of differential and integral calculus. The discussion is confined to one- and two-dimensional space, but even so, this requires a knowledge of both ordinary and partial differential equations. The skills emphasized here are first and most importantly the translation of geologic processes or systems into dynamical models. By *dynamical model* we mean a physical–mathematical description of changes in important geological variables, such as dissolution of a mineral or variations in thickness of river deposits. It is this translation process that we want to emphasize. It lies at the core of what scientists do, whereas one can always go to the math department for a solution once nature has been abstracted into mathematical form. Having said that, it often is very frustrating not knowing immediately what the solution space "means" in terms of the problem. Consequently, we provide some instruction in obtaining numerical solutions to sets of differential equations that have been transformed into finite-difference approximations. Finally, we show how numerical experiments enhance our geological understanding.

We call this book a primer because it introduces the reader to modeling of dynamical systems of continuous variables but does not cover all fields in the geosciences nor all methods. Readers looking for a broader range of disciplines might turn to *Mathematical Models in the Applied Sciences* by A. C. Fowler (1997) and for a broader range in types of modeling turn to *Mathematical Modelling for Earth Sciences* by Xin-She Yang (2008). More disciplinary-specific and in-depth treatments are offered in *Fluid Physics in Geology* by D. J. Furbish (1997), *Quantitative Modeling of Earth Surface Processes* by Jon Pelletier (2008), *The Mechanics and Chemistry of Landscapes* by R. S. Anderson and S. P. Anderson (2010), *Numerical Adventures with Geochemical Cycles* by J.C.G. Walker (1991), *Diagenetic Models and Their Interpretation* by B. P. Boudreau (1997), and *Numerical Methods in the Hydrological Sciences* by G. Hornberger and P. Wiberg (2005). For more in-depth discussions of finite difference techniques, the reader is referred to *Computational Fluid Dynamics for Engineers* by K. A. Hoffmann and S. T. Chiang (2000) and *Computational Techniques for Fluid Dynamics* by C.A.J. Fletcher (1991).

MATHEMATICAL MODELING
of Earth's Dynamical Systems

Modeling and Mathematical Concepts

> A system is a big black box
> Of which we can't unlock the locks,
> And all we can find out about
> Is what goes in and what comes out.
> —*Kenneth Boulding*

Kenneth Boulding—presumably somewhat tongue-in-cheek—expresses the cynic's view of systems. But this description will only be true if we fail as modelers, because the whole point of models is to provide illumination; that is, to give insight into the connections and processes of a system that otherwise seems like a big black box. So we turn this view around and say that Earth's systems may each be a black box, but a well-formulated model is the key that lets you unlock the locks and peer inside.

There are many different types of models. Some are purely conceptual, some are physical models such as in flumes and chemical experiments in the lab, some are stochastic or structure-imitating, and some are deterministic or process-imitating. The distinction also can be made between forward models, which project the final state of a system, and inverse models, which take a solution and attempt to determine the initial and boundary conditions that gave rise to it. All of the models described in this book are deterministic, forward models using variables that are continuous in time and space. One should think of the models as physical–mathematical descriptions of temporal and/or spatial changes in important geological variables, as derived from accepted laws, theories, and empirical relationships. They are "devices that mirror nature by embodying empirical knowledge in forms that permit (quantitative) inferences to be derived from them" (Dutton, 1987). The model descriptors are the conservation laws, laws of hydraulics, and first-order rate laws for material fluxes that predict future states of a system from initial conditions (ICs), boundary conditions (BCs), and a set of rules. For a given set of BCs and ICs, the model will always "determine" the same final state. Furthermore, these models are mathematical (numerical). We emphasize this type of model over other types because it represents a large proportion of extant models in the earth sciences. Dynamical models also provide a good vehicle for teaching the art of modeling. We call modeling an art because one must know what one wants out of a model and how to get it. Properly constructed, a model will rationalize the information coming to our senses, tell us what the most important data are, and tell us what data will best test our notion of how nature works as it is embodied in the model. Bad models are too complex and too uneconomical or, in other cases, too simple.

Pros and Cons of Dynamical Models

The advantage of a deterministic dynamical model is that it states formal assertions in logical terms and uses the logic of mathematics to get beyond intuition. The logic is as follows: If my premises are true, and the math is true, then the solutions must be true. Suddenly, you have gotten to a position that your intuition doesn't believe, and if upon further inspection, your intuition is taught something, then science has happened. Models also permit formulation of hypotheses for testing and help make evident complex outcomes, nonlinear couplings, and distant feedbacks. This has been one of the more significant outcomes of climate modeling, for example. If there are leads and lags in the system, it's tough for empiricists because they look for correlation in time to determine causation. But if it takes a couple of hundred years for the effect to be realized, then the empiricist is often thwarted.

Particularly relevant for geoscientists and astrophysicists, dynamical models also permit controlled experimentation by compressing geologic time. Consider the problem of understanding the collision of galaxies—how does one study that process? Astrophysicists substitute space for time by taking photographs of different galaxies at different stages of collision and then assume they can assemble these into a single sequence representing one collision. That sequence acts as a data set against which a model of collision processes can be tested where the many millions of years are compressed. The idea of a snowball Earth provides an example even closer to home, or one could ask the question: What did rivers in the earthscape look like prior to vegetation? Questions of this sort naturally lend themselves to idea-testing through dynamical models.

But dynamical models not properly constructed or interpreted can cause great trouble. Recently, Pilkey and Pilkey-Jarvis (2007) passionately argued that many environmental models are not only useless but also dangerous because they have made bad predictions that have led to bad decisions. They argue that there are many causes, including inadequate transport laws, poorly constrained coefficients ("fudge factors"), and feedbacks so complex that not even the model developers understand their behavior. Although we think the authors have painted with too broad a brush, we agree with them on one point. A simple falsifiable model that has been properly validated [even if in a more limited sense than that of Oreskes et al. (1994)] is better than an ill-conceived complex model with scores of poorly constrained proportionality constants [also see Murray (2007) for a discussion of this point]. Finally, we should never lose sight of the fact that in a model "it is not possible simultaneously to maximize generality, realism, and precision" (atmospheric scientist John Dutton, personal communication, 1982).

An Important Modeling Assumption

We assume in this book that a fruitful way to describe the earth is a series of mathematical equations. But is this mathematical abstraction an adequate description of reality? Does reality exist in our minds as mathematical formulas or is it outside of us somewhere? For example, the current understanding of the fundamental physical laws that govern the universe—string theory—is entirely a mathematical theory without experimental confirmation. To some it unites the general theory of relativity and quantum mechanics into a final unified theory. To others it is unfalsifiable and infertile (see, e.g., Smolin, 2006).

We avoid these philosophical problems by simply asserting that mathematical descriptions of the earth both past and present have proved to be a useful way of knowing. As the Nobel Laureate Eugene Wigner noted, "The miracle of the appropriateness of the language of mathematics for the formulation of the laws of physics is a wonderful gift which we neither understand nor deserve" (Wigner, 1960). An alternative view is that they are inherently quite limited in their predictive power. This view is summarized cogently by Chris Paola in a review of sedimentary models: "[A]ttempting to extract the dynamics at higher levels from comprehensive modelling of everything going on at lower levels is . . . like analyzing the creation of La Boheme as a neurochemistry problem" (Paola, 2000). Whereas we accept this point of view in the limit, we reject it for a wide range of complex systems that are amenable to reduction.

Some Examples

To set the stage for the chapters that follow, we present two problems for which modeling can provide insight. Other examples abound in the literature. Of special note for those studying Earth surface processes is the Web site of the Community Earth Surface Dynamics Modeling Initiative (CSDMS; pronounced "systems"). CSDMS (http://

csdms.colorado.edu) is a National Science Foundation (NSF)-sponsored community effort providing cyberinfrastructure aiding the development and dissemination of models that predict the flux of water, sediment, and solutes across the earth's surface. There one can find hundreds of models that incorporate the conservation and geomorphic transport laws and that can be used to solve particular problems. A companion organization, Computational Infrastructure for Geodynamics (http://www.geodynamics .org/), provides similar support for computational geophysics and related fields.

Example I: Simulation of Chicxulub Impact and Its Consequences

Probably *the* most famous event in historical geology, at least from the public's perspective, is the extraterrestrial impact event at the end of the Mesozoic Era that killed off the dinosaurs. Most schoolchildren know the standard story: A large asteroid that struck the surface of the earth in Mexico's Yucatán Peninsula created the Chicxulub Crater along with a rain of molten rock, toxic chemicals, and sun-obscuring debris that eliminated roughly three-quarters of the species living at the time. To work through the specific details of what happened and to predict the consequences of such an uncommon event is not easy because the physical and chemical processes are operating in a pressure–temperature state all but impossible to obtain experimentally. It is precisely these cases that benefit most from numerical simulation.

But is an asteroid impact computable? That is, given as many conservation equations and rate laws as there are state variables, and given initial and boundary conditions, can future states of the system be predicted with an acceptable degree of accuracy? Gisler et al. (2004) thought so. They derived a model simulating a 10-km-diameter iron asteroid plunging into 5 km of water that overlays 3 km of calcite, 7 km of basalt crust, and 6 km of mantle material. The set of equations was solved using the SAGE code from Los Alamos National Laboratory and the Science

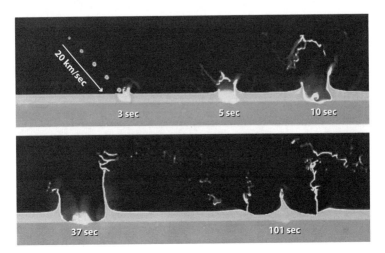

Figure 1.1. Montage of images from a three-dimensional (3-D) simulation of the impact of a 1-km-diameter iron bolide at an angle of 45 degrees into a 5-km-deep ocean. Maximum transient crater diameter of 25 km is achieved at about 35 seconds. [From Gisler, G. R., et al. (2004). Two- and three-dimensional asteroid impact simulations. *Computing in Science & Engineering* 6(3):46–55. Copyright © 2004 IEEE. Reproduced with permission.]

Applications International Corporation, which was developed under the U.S. Department of Energy's program in Accelerated Strategic Computing. Their model contained 333 million computational cells and used 1,024 processors for a total computational time of 1,000,000 CPU hours on a cluster of HP/Compaq PCs.

The results (fig. 1.1) document the dissipation of the asteroid's kinetic energy (which amounts to about 300 teratons TNT equivalent, or ~4×10^{21} J). The impact produces a tremendous explosion that melts, vaporizes, and ejects a substantial volume of calcite, granite, and water. Predictions from the model aid in understanding how, why, and where the resulting environmental changes caused the extinction.

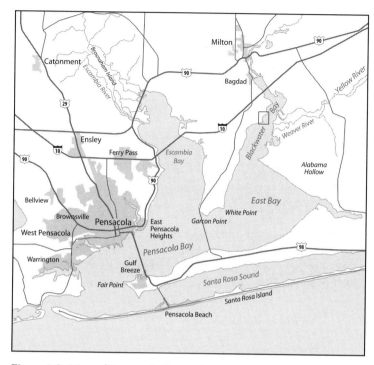

Figure 1.2. Map of Pensacola Bay and surrounding area. Hurricane Ivan passed on a trajectory due north just 20 mi to the west. The rectangle drawn in Blackwater Bay encompasses the region of interest. (Map adapted from a U.S. Geological Survey 1:250,000 topographic map.)

Example II: Storm Surge of Hurricane Ivan in Escambia Bay

On September 16, 2004, Hurricane Ivan made landfall about 35 mi (56 km) west of Pensacola, Florida (fig. 1.2). At the time of landfall, peak winds exceeded 125 mi h^{-1} (200 km h^{-1}), severely damaging many buildings in the Pensacola area. Probably equally damaging, however, was the surge of water along the coast and up Pensacola Bay. Homeowners along the bay experienced significant

Figure 1.3. The scene 3 hours after the eye passes. (Photo courtesy of Ray Slingerland.)

flooding (fig. 1.3) even though some were more than 25 mi by water from the open ocean.

Was this event an unpredictable act of God or could we have predicted the flooding? As you might suspect, the answer is that not only could it have been predicted, it was (fig. 1.4).

In chapter 10, we describe how surge models of the sort used by the U.S. Army Corps of Engineers are derived.

Steps in Model Building

So how does one construct a model of a geological phenomenon? Throughout this book, we will try to follow some logical steps in model development. First, get the physical picture clearly in mind. As an example, say one wanted to model the number of flies in a room as a function of time. The physical picture includes defining the

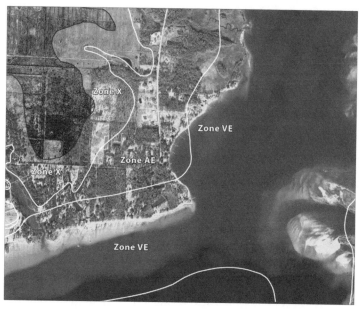

Figure 1.4. Observed surge high-water line (solid gray) versus those predicted (solid white) for Hurricane Ivan. Zone VE: Area subject to inundation by the 1%-annual-chance flood event with additional hazards due to storm-induced velocity wave action. Zone AE: Area subject to inundation by the 1%-annual-chance flood event determined by detailed methods. Zone X: Area of minimal flood hazard higher than the elevation of the 0.2%-annual-chance flood. See figure 1.2 for location. (From http://www.fema.gov/pdf/hazard/flood/recovery data/ivan/maps/K33.pdf.)

dependent variable(s) (in this case the number of flies), the independent variables (time), and the size of the room. Second, one must define the physical processes to be treated and the boundaries of the model. The processes in the case of flies are flying, crawling, hatching, and dying. The boundaries of the model are those that do not pass flies such as walls, floor, and ceiling, and open boundaries such as doors and windows. Third, write down the physical laws to be used. Generally, these will be laws such as conservation of mass, Fick's law, and so on. In the case of flies, the laws are rate laws governing the flux of flies into and out of the room and laws defining the rates at which flies are created and die within the room. Fourth, put down very clearly the restrictive assumptions made. If one assumes that the flies will enter the room in proportion to the gradient in their number between inside and outside, write that assumption down. Fifth, perform a balance, first in words and then in symbols. Usually, one balances properties such as force, mass, or number. In the case of flies, we would say

The time rate of change of flies in the room
  = the rate at which they enter through doors and
    windows
    – the rate at which they leave
      + the rate at which they are born
        – the rate at which they die.

We would then substitute symbols for number of flies, time, and so forth. Sixth, check units. All the terms in the balance equation must be of the same units; if they are not, we have made a mistake in our definitions, and now is the time to catch it. Seventh, write down initial and boundary conditions. By initial conditions are meant the values of the dependent variables at the start of the calculations. For example, we would specify the number of flies in the room at $t = 0$ as zero or some finite number. Boundary conditions are the values of the dependent variables at the edges of the spatial domain of interest. For example, we must specify the number of flies outside as a function of time and specific door or window. Lastly, solve the mathematical model. If you are lucky you can find an equation of similar form that has already been analytically solved. There is value in pursuing an analytic solution even if you need to reduce variable coefficients to constants or even drop terms, because the simplified equation will provide insight into your system's behavior. But often no analytic solutions will be available, and this step will require converting the equation set into a numerical form amenable for solution on a computer. Finally, you should verify and

Table 1.1. Steps in Problem Solving

1. Get the physical picture clearly in mind.
2. Define the physical processes to be treated and the boundaries of the model.
3. Write down the laws and transport functions to be used.
4. Put down very clearly the restrictive assumptions made.
5. Perform the balance, first in words and then in symbols.
6. Check units.
7. Write down initial and boundary conditions.
8. Verify, validate, and solve the mathematical model.

validate your model. According to Oberkampf and Trucano (2002), verification is the process of determining that a model implementation accurately represents your conceptual description of the model and the solution to the model. Thus, verification checks that the coding correctly implements the equations and models, whereas validation determines the degree to which a model is an accurate representation of the real world from the perspective of its intended uses. In other words, does the model agree with reality as observed in experiments and in the field. To formalize your thinking as you approach a problem, follow all of these steps in table 1.1.

Basic Definitions and Concepts

Why Models Are Often Sets of Differential Equations

We naturally find it easier to think about how an entity changes than about the entity itself. For example, my car speedometer measures my velocity, not the distance I've traveled from my garage since I started my trip. It is easier to state that the time rate of change of water in my boat equals the rate at which water enters through the open seams minus the rate at which I am bailing it out than it is to state how the volume of water actually varies with time. Changing entities of this sort are called variables, of which there are two kinds: independent (space and time) and dependent, by which we mean the state variables in question (velocity, mass of water, and so forth). The rate of change of one variable with respect to another is called a derivative, written, for example, as the ordinary derivative dV/dt if the dependent variable V is only a function of the independent variable t, or the partial derivative $\partial V/\partial t$ if V also depends upon other independent variables. Equations that express a relationship among these variables and their derivatives are differential equations.

However, often we want to know how the variables are related among themselves, not how they are related to their derivatives. So the general procedure is to derive the differential equations from first principles and then solve them for the values of the dependent variables as functions of the independent variables and other parameters.

To solve the differential equations requires more than the differential equation itself, however. The problem must be *well posed*. A well-posed problem contains as many governing equations as there are dependent variables. Also, the time and space interval over which the solution is to be obtained should be specified, and additional information concerning the dependent variables must be supplied at the start time (called initial conditions, or ICs) and the boundaries of the intervals (called boundary conditions, or BCs). This information is necessary because integration of the differential equations creates constants of integration in the case of ordinary differential equations (ODEs) and functions of integration in the case of partial differential equations (PDEs). The number of constants or functions needed is equal to the order of the differential equation. Thus, for a partial differential equation that is second order in both time and space, one must supply two functions derived from the ICs specifying the dependent variable as a function of time and two functions derived from the BCs specifying the dependent variable as a function of space. There are three possible types of BC information that can be supplied.

Dirichlet Conditions

In this type of BC, the solution itself is prescribed along the boundary, as, for example, if we were to set dependent variable $C(0,t) = P$, where P is some temporally constant value of the dependent variable.

Neumann Conditions

Alternatively, the derivatives of the solution in the normal direction to the boundary are prescribed. For any variable that obeys a first-order rate law, this is equivalent to specifying the flux across the boundary. Thus, we might know that a chemical species of concentration $C(x,t)$ diffuses across a plane in an aquifer at $x = 0$ at a flux $q = q_0$, and therefore the BC at $x = 0$ becomes

$$D\frac{\partial C}{\partial x}\bigg|_{x=0} = -q_o. \tag{1.1}$$

Mixed Conditions

This BC, sometimes called a "Robin" boundary condition, combines both of the above types. For example, if the flux through the face at $x = 0$ was not constant, but was proportional to the difference between a fixed concentration A at $x = -1$ and $C(0,t)$, the actual concentration at $x = 0$, then the appropriate BC would be

$$D\frac{\partial C}{\partial x}\bigg|_{x=0} = -k[A - C(0,t)], \tag{1.2}$$

where k is a proportionality constant with units of m s^{-1}.

Finally, for a well-posed problem, a solution must exist, be unique, and depend continuously on the auxiliary data. Most geoscience problems have solutions, and most can be made unique with proper BCs, although one should be aware that underprescription of BCs leads to nonuniqueness. The third requirement is met when small changes in BCs lead to small changes in the solution.

Nondimensionalization

Before attempting a solution, it is always useful to rewrite the well-posed problem using nondimensional variables (see table 1.2). When we nondimensionalize equations, we remove units by a suitable substitution of variables. This

Table 1.2. Steps in Nondimensionalization

1. Identify all the independent and dependent variables.
2. Define a nondimensional term for each variable by scaling each variable with a coefficient in the problem with the same units.
3. Substitute each definition into the governing equation and divide through by the coefficient of the highest-order polynomial or derivative.
4. If you have chosen well, the coefficients of many terms will become 1.

process groups together various coefficients into ensembles called parameters, thereby allowing us to predict natural system behavior more easily. We also can describe the solution in terms of a few parameters composed of the various dimensional geometric and material properties in the problem. Sometimes characteristic properties of a system emerge from these, such as a resonance frequency. Plus, one solution fits all; we don't need to define a new solution if we want to change a parameter. Finally, if we have chosen well, the solutions scale between 0 and 1, thereby allowing us to better control accuracy if the solution must be obtained by numerical techniques. The nondimensionalization process, also known as scaling, will be illustrated in detail after we have created some models.

A Brief Mathematical Review

Here we review some mathematical concepts used in the creation and solution of well-posed dynamical models. We usually seek a solution over a portion or interval of time and space. An interval is formally defined as the set of all real numbers between any two points on the number line of space or time and will be denoted as: $a < x < b$.

Definition of a Function

If to each value of an independent variable x in a specified interval there is one and only one real value of the dependent variable y, then y is a *function* of x in the interval. The concept can be extended to functions of n independent variables. For example,

$$z = f(x,y) = x + y. \tag{1.3}$$

There are two types of functions: explicit and implicit. The relationship $f(x,y) = 0$ defines y as an implicit function of x. Implicit solutions of equations often are pointless, as, for example,

$$f(x,y) = x^3 + y^3 - 3xy = 0, \tag{1.4}$$

which still does not tell us explicitly the value of y for a given x.

Ordinary Differential Equation

Let $f(x)$ define a function of x on an interval. By ordinary differential equation (ODE) we mean an equation involving x, the function $f(x)$, and its derivatives. The order is the order of the highest derivative. For any function $y = f(x)$, the geometrical meaning of the first derivative is the slope of the line tangent to a point on the function, and the second derivative is the curvature of the function at that point.

Solution of an Ordinary Differential Equation

Let $y = f(x)$ define y on an interval. $f(x)$ is an explicit solution if it satisfies the equation for *every* x on the interval, or if upon substitution, the ODE reduces to an identity.

Fundamental Theorem of Calculus

Integration is antidifferentiation. Thus, if:

$$y = x^2$$

and

$$\frac{dy}{dx} = 2x$$

then

$$\int dy = \int 2x\,dx = x^2 + c, \tag{1.5}$$

where c is a constant of integration.

General Solution

For a very large class of ODEs, the solution of an ODE of order n contains n arbitrary constants. Example:

$$\frac{d^2y}{dx^2} = x$$
$$y = \frac{x^3}{6} + c_1 x + c_2. \tag{1.6}$$

The n-parameter family of solutions, $y = f(x, c_1 \dots c_n)$, to an nth order ODE is called a *general solution*. Constants are called constants of integration. To find a particular solution requires additional information to uniquely specify the constant(s). This additional information comes from the initial or boundary conditions.

Systems of Ordinary Differential Equations

The pair of equations

$$\frac{dx}{dt} = f_1(x,y,t)$$
$$\frac{dy}{dt} = f_2(x,y,t) \tag{1.7}$$

is called a system of two first-order ODEs. A *solution* is then a pair of functions $x(t)$, $y(t)$ on a common interval of t.

The Partial Derivative

If $z = f(x,y)$, then the partial derivative of z with respect to x at (x,y) is

$$\frac{\partial z}{\partial x} = \lim_{h \to 0} \frac{f(x+h,y) - f(x,y)}{h}, \tag{1.8}$$

and so forth. Note that a partial with respect to x is differentiated with y being regarded as a constant. The geometrical interpretation of the partial derivative is given in figure 1.5. Geologists will recognize that the solution surface at a point may be characterized by two apparent dips, one in the x direction and one in the y direction. These

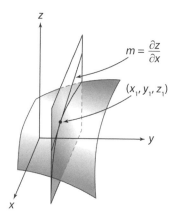

Figure 1.5. Geometrical meaning of a partial derivative. The curved surface is the value of z as a function of x and y. A line drawn tangent to the surface in the x,z plane at position y_1 has slope m equivalent to the value of the partial derivative at (x_1,y_1).

slopes are given by the partial derivatives, and therefore the apparent dip angles are given by the arctangents of the partial derivatives.

Differential of a Function of Two Independent Variables

If $z = f(x,y)$, then the differential of z is

$$dz = \frac{\partial f(x,y)\,dx}{\partial x} + \frac{\partial f(x,y)\,dy}{\partial y}. \tag{1.9}$$

Partial Differential Equations

An equation involving two or more independent variables, x_i, the function, $f(x_i)$, and its partial derivatives is called a partial differential equation (PDE). The order is the order of the highest partial derivative.

It is always helpful to classify the PDEs of your problem, because much can be learned about the behavior of the solution even without obtaining the actual solution. In fact, the method of solution often is class-dependent. A PDE can be linear or nonlinear, with the nonlinear equations being more difficult to solve. A PDE is linear if the dependent variable and all its derivatives appear in a linear fashion; that is, are not multiplied by each other, squared, and so forth. It is homogeneous if it lacks a term that is independent of the dependent variable.

Kinds of Coefficients

Coefficients may be constants, functions of the independent variables, or functions of the dependent variables. In the latter case, the equation is said to be nonlinear.

Three Basic Types of Linear Partial Differential Equations

A second-order linear equation in two variables is of the form

$$A\frac{\partial^2 u}{\partial x^2} + B\frac{\partial^2 u}{\partial x \partial y} + C\frac{\partial^2 u}{\partial y^2} + D\frac{\partial u}{\partial x} + E\frac{\partial u}{\partial y} + F = 0, \quad (1.10)$$

where A through F are constants or functions of x and y. All linear equations like equation 1.10 can be classified according to the following scheme. If:

$$B^2 - 4AC < 0 \Rightarrow \text{The PDE is elliptic:}$$
$$B^2 - 4AC = 0 \Rightarrow \text{The PDE is parabolic:}$$
$$B^2 - 4AC > 0 \Rightarrow \text{The PDE is hyperbolic:} \quad (1.11)$$

The usefulness of this classification will be shown later.

Solution of a Partial Differential Equation

A function $z = f[x,y,g_i(x,y)]$, is a solution if it satisfies the PDE upon substitution. Note that PDEs of the nth order require n *functions* of integration, $g_i(x,y)$.

Chain Rule

Suppose $z = f(x,y)$, and $x = F(t)$, and $y = G(t)$ where F and G are functions of t. What is dz/dt? Because z is a function of x and y, and x and y are functions of t:

$$\frac{dz}{dt} = \frac{\partial z}{\partial x}\frac{dx}{dt} + \frac{\partial z}{\partial y}\frac{dy}{dt}. \quad (1.12)$$

Product Rule

If $f(x) = u(x)\,v(x)$, then

$$\frac{\partial f}{\partial x} = u\frac{\partial v}{\partial x} + v\frac{\partial u}{\partial x}. \quad (1.13)$$

Taylor Series Expansion

Taylor's theorem was first derived by Brook Taylor, who was born August 18, 1685, in Edmonton, Middlesex, England. Its importance remained unrecognized until 1772 when Lagrange proclaimed it the basic principle of the differential calculus. Taylor showed that if one knows the value of a function at (x,y), then the value of the function at $(x + dx, y)$ can be approximated as

$$
f(x + dx, y) = f(x,y) + \frac{1}{1!}\frac{\partial f(x,y)}{\partial x}dx
$$
$$
+ \frac{1}{2!}\frac{\partial^2 f(x,y)}{\partial x^2}(dx)^2 + \ldots\ldots, \tag{1.14}
$$

where the ellipses denote all higher-order terms in the series.

Substantial Time Derivative

Let us say we are interested in the rate at which the temperature, T, changes as we drive south in the winter from Pennsylvania to Florida. We recognize that there will be two sources of temperature change: one arising due to the change of temperature independent of any change in location (say the normal heating that occurs as night turns to day), and one arising because we are moving south through the latitudinal temperature gradient at our car speed u. Equation 1.12 captures this idea. Let $z = T$, $x =$ distance $= F(t)$, and $y = G(t) = t$. Therefore, the total time derivative of T is

$$
\frac{dT}{dt} = \frac{\partial T}{\partial x}\frac{dx}{dt} + \frac{\partial T}{\partial t}\frac{dy}{dt}. \tag{1.15}
$$

However, $dx/dt = u$, the car velocity, and $dy/dt = 1$; therefore,

$$
\frac{dT}{dt} = u\frac{\partial T}{\partial x} + \frac{\partial T}{\partial t}. \tag{1.16}
$$

The righthand side (RHS) of equation 1.16 is called the substantial time derivative (in this case in only one dimension) and often written in shorthand form as DF/Dt, where F is the dependent variable in question.

Concept of a Control Volume

A control volume is the region of space we define to perform a balance of mass, energy, and so forth. It can be either macroscopic, such as a finite volume of a river channel, or microscopic with dimensions dx, dy, dz, for example. Choosing the control volume for a problem is somewhat an art. Ideally, the boundaries should be meaningful physical surfaces through which fluxes can be easily specified without recourse to complicated geometric formulas.

The Basic Scientific Laws, Axioms, and Definitions

All of the physics and chemistry used in this book can be reduced to only 17 basic concepts. These are listed in table 1.3 for later reference.

Table 1.3. Basic Laws, Axioms, and Definitions

I. Conservation of Mass
The time rate of change of mass in a control volume equals the mass rate into the volume minus the mass rate out.

II. Newton's First Law
Any body is in a state of rest or in uniform rectilinear motion until some forces applied to it produce a change in the state of the body (motion or deformation). (NB: body = discrete entity with mass.)

III. Newton's Second Law
The rate of change of momentum of a body is proportional to the impressed force and is made in the direction of the straight line in which the force is impressed.

IV. Newton's Third Law
To every action there is always opposed and equal reaction, or the mutual actions of two bodies upon each other are always equal in magnitude and opposite in direction.

V. Corollary I
A body acted on by two forces simultaneously will move along the diagonal of a parallelogram in the same time as it would move along the sides by those forces acting separately.

VI. Conservation of Momentum
Using Newton's third law to extend Newton's second law to the total momentum of systems of particles:
The time rate of change of momentum in a control volume equals the time rate in of momentum minus the time rate out plus the sum of forces.

(continued)

Table 1.3. (*continued*)

VII. The Coriolis Force
Arises out of a choice to apply the laws of motion developed for an inertial reference frame to a rotating reference frame that is attached to Earth. It is quantified as twice the product of the angular velocity and the sine of the latitude.

VIII. Quadratic Drag Law
The force experienced by a large object moving through a fluid at relatively large velocity (i.e., with a Reynolds number greater than ~1,000) is proportional to the square of the velocity.

IX. Universal Law of Gravitation
Between any two particles of mass m_1 and m_2 at separation R, there exist attractive forces F_{12} and F_{21} directed from one body to the other and equal in magnitude to the product of masses and inversely proportional to square of distance between them.

X. Equivalence of Work and Energy
Work is measured by the product of an acting force and the distance traveled by a body. It is a measure of the transfer of energy from one body to another.

XI. Conservation of Energy
Energy retains a constant value in all the changes of the form of motion.

XII. Stefan–Boltzmann Law
Energy radiated from a black body is proportional to the fourth power of its temperature (Kelvin units).

XIII. First-Order Rate Laws
A substance flows down a potential or concentration gradient at a rate proportional to the magnitude of the gradient. Includes Fourier's law, Darcy's law, Newton's law of viscosity, Ohm's law, Hooke's law, and Fick's first law.

XIV. Law of Mass Action
The rate of a forward chemical reaction is proportional to the product of the reactants' concentrations (raised to the power of their stoichiometric coefficients).

XV. Law of Radioactive Decay
The rate of decay of a radioactive substance is proportional to its mass.

XVI. Relationship Between Stress and Strain
The shear stress acting on a Newtonian fluid is proportional to the rate of shear strain, with the proportionality constant being the coefficient of viscosity.

XVII. Archimedes' Principle
A body partly or wholly immersed in a fluid is buoyed up by a force acting vertically upwards through the center of mass of displaced fluid and equal to the weight of the fluid displaced.

Summary

This chapter was designed to instill in the reader a sense of the role of mathematical models in the geosciences, especially those that we focus on here—dynamical systems models. Following on some examples, we have provided a template for constructing mathematical models that we will follow religiously in this book. Many of the terms and concepts that we use in later chapters were introduced, and some necessary basic mathematics was reviewed for those needing a reminder. Now that the toolbox has been filled, we move on to the process of converting differential equations into algebraic expressions that can be solved using matrix algebra: the process of obtaining numerical solutions by finite difference.

Basics of Numerical Solutions by Finite Difference

Some models give rise to relatively simple analytic solutions for a wide range of initial and boundary conditions. But this is generally not true for more complex partial differential equations of increased dimension. As the dimensions and complexity of the coefficients and boundary conditions increase, finding analytic solutions becomes prohibitively difficult, and, in fact, some nonlinear PDEs have no known analytic solutions. To circumvent this problem, numerical solution schemes have been developed that involve finding discrete solutions at specific points in time and space. Of these schemes, the simplest are of the finite difference type, and we restrict our discussion to them. In essence, the approach is to convert the differential equations of a well-posed problem into a set of linear algebraic equations written in terms of the dependent variable(s). Because matrix algebra plays such a large role in solving such sets, we first briefly review it here. For a more complete introduction to discretization and finite difference methods, the reader is referred to Fletcher (1991) and Hoffman and Chiang (2000).

First Some Matrix Algebra

A matrix is a table or array of numbers or algebraic variables arranged in rows and columns such as this table for matrix **A**:

$$\mathbf{A} = \begin{bmatrix} a_{1,1} & a_{1,2} & a_{1,3} \\ a_{2,1} & a_{2,2} & a_{2,3} \\ a_{3,1} & a_{3,2} & a_{3,3} \end{bmatrix}. \tag{2.1}$$

Matrices are usually denoted in bold. The elements or entries in the matrix are usually denoted by indices reflecting the row and column of the entry, with the row number first. A matrix such as equation 2.1 having three rows and three columns is said to be a 3 by 3, or 3×3, matrix. If the number of rows and columns is equal, then the matrix is said to be square. Let the number of rows be m and the number of columns be n. Then if $m = 3$ and $n = 1$, such as

$$\mathbf{u} = \begin{pmatrix} a \\ b \\ c \end{pmatrix}, \tag{2.2}$$

matrix $\mathbf{u}$ is said to be a column matrix or column vector.

Matrices can be added or subtracted only if their number of rows and number of columns is identical, in which case corresponding entries are added or subtracted. Two matrices $\mathbf{A}$ and $\mathbf{B}$ can be multiplied only if the number of columns of $\mathbf{A}$ is equal to the number of rows of $\mathbf{B}$. For example, if matrix $\mathbf{A}$ has size m by n, then it may premultiply a matrix $\mathbf{B}$ with size n by q, in which case the product matrix $\mathbf{C} = \mathbf{AB}$ will be size m by q. In component form, this multiplication takes the form

$$C_{ij} = \sum_{k=1}^{n} a_{ik} b_{kj}. \tag{2.3}$$

A scalar multiplies a matrix by multiplying each entry. Likewise, differentiation of a matrix is done on each element.

A diagonal matrix is a square matrix in which $a_{ij} = 0$ if i does not equal j. For example, this is a special diagonal matrix called the identity matrix $\mathbf{I}$:

$$\mathbf{I} = \begin{bmatrix} 1 & 0 & 0 \\ 0 & 1 & 0 \\ 0 & 0 & 1 \end{bmatrix}. \tag{2.4}$$

The transpose of a matrix (call it matrix $\mathbf{B}$), denoted by $\mathbf{B}^{\mathrm{T}}$, is obtained by reflecting the entries about the diagonal from upper left to lower right.

The inverse of a square matrix $\mathbf{A}$ is defined such that

$$\mathbf{A}^{-1}\mathbf{A} \equiv \mathbf{A}\mathbf{A}^{-1} = \mathbf{I} \tag{2.5}$$

and is denoted $\mathbf{A}^{-1}$. A matrix is invertible if and only if its determinant is nonzero. The determinant is a special number that can be computed for any square matrix $\mathbf{A}$ and is denoted $\det(\mathbf{A})$ or $|\mathbf{A}|$. Computing the determinant depends upon the dimensions of the matrix. For example, the determinant of the 3×3 matrix

$$\mathbf{A} = \begin{bmatrix} a & b & c \\ d & e & f \\ g & h & i \end{bmatrix} \tag{2.6}$$

is given by: $\det(\mathbf{A}) = aei - afh + bfg - bdi + cdh - ceg$.

Solution of Linear Systems of Algebraic Equations

Consider the system of m algebraic equations in n unknowns:

$$\begin{aligned} a_{11}x_1 + a_{12}x_2 + \cdots + a_{1n}x_n &= b_1 \\ a_{21}x_1 + a_{22}x_2 + \cdots + a_{2n}x_n &= b_2 \\ \vdots \qquad\qquad \vdots \\ a_{m1}x_1 + a_{m2}x_2 + \cdots + a_{mn}x_n &= b_m. \end{aligned} \tag{2.7}$$

Using the rules noted above, one can write this system of equations in the compact notation of matrix algebra as

$$\begin{pmatrix} a_{11} & a_{12} & \cdots & a_{1n} \\ a_{21} & a_{22} & \cdots & a_{2n} \\ \vdots & \vdots & \ddots & \vdots \\ a_{m1} & a_{m2} & \cdots & a_{mn} \end{pmatrix} \begin{pmatrix} x_1 \\ x_2 \\ \vdots \\ x_n \end{pmatrix} = \begin{pmatrix} b_1 \\ b_2 \\ \vdots \\ b_m \end{pmatrix} \tag{2.8}$$

or

$$\mathbf{A}x = \mathbf{b}. \tag{2.9}$$

Usually, we want to solve for the column vector x of unknowns. Because it will always be the case that $m = n$ in these systems, we can use the definition from equation 2.5 to obtain

$$\mathbf{A}^{-1}\mathbf{A}x = \mathbf{A}^{-1}\mathbf{b}$$

or

$$x = \mathbf{A}^{-1}\mathbf{b}. \tag{2.10}$$

General Finite Difference Approach

To introduce the basics of the finite difference technique, consider the generic partial differential equation representing one-dimensional diffusion:

$$\frac{\partial T}{\partial t} - D\frac{\partial^2 T}{\partial x^2} = 0. \tag{2.11}$$

At this point, it is not necessary to know how the equation is derived, nor even what property T represents, only that $T(x,t)$ is a continuous function. To solve for $T(x,t)$ over specific intervals of time and distance and for specific initial and boundary conditions using the finite difference method, the general approach is to rewrite equation 2.11 as an algebraic equation and solve that equation at discrete points in space and time. Figure 2.1 shows the steps. We begin by discretizing the x–t plane.

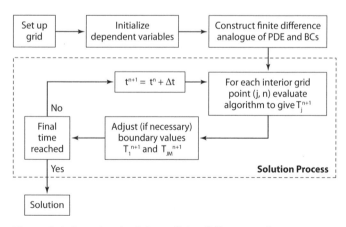

Figure 2.1. Steps in obtaining a finite difference solution to a PDE. [Modified from Fletcher, C.A.J. (1991). *Computational Techniques for Fluid Dynamics.* Berlin, Springer-Verlag.]

Discretization

Consider the domain in x–t space in figure 2.2 as the region in which we seek a solution to equation 2.11. In reality the solution is a continuous surface, but in numerical solutions the space–time plane is discretized into a set of points. The points are not of necessity at equal intervals, but for simplicity here we take the space step as a constant Δx and the time step as a constant Δt. Thus points in x and t lie at $x = j\Delta x$ and $n\Delta t$ where $j = 1, 2, 3, \ldots JM$ (maximum value of j) and $n = 1, 2, 3, \ldots NM$ (maximum value of n). We seek the values of the solution only at these discrete points. Of course if Δx and Δt are very small, then the coverage of solutions approximates the continuous solution surface. For an in-depth discussion of discretization for structured grids, see Hoffmann and Chiang (2000) or Anderson (1995).

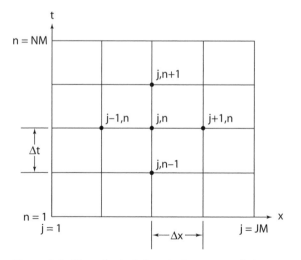

Figure 2.2. Hypothetical domain in space and time within which solutions of the one-dimensional diffusion equation are sought.

Obtaining Difference Operators by Taylor Series

To obtain an algebraic equation representing equation 2.11, we use the Taylor series. Consider a function $u(x,t)$. Equation 2.12 estimates the value of a function u at a point Δx ahead of the point x where the function is known, and equation 2.13 estimates the function at a point one space step behind:

$$u(x + \Delta x) = u(x) + \Delta x \frac{\partial u}{\partial x} + \frac{\Delta x^2}{2} \frac{\partial^2 u}{\partial x^2}$$
$$+ \frac{\Delta x^3}{6} \frac{\partial^3 u}{\partial x^3} + O(\Delta x^4) \tag{2.12}$$

$$u(x - \Delta x) = u(x) - \Delta x \frac{\partial u}{\partial x} + \frac{\Delta x^2}{2} \frac{\partial^2 u}{\partial x^2}$$
$$- \frac{\Delta x^3}{6} \frac{\partial^3 u}{\partial x^3} + O(\Delta x^4). \tag{2.13}$$

The term $O(\Delta x)^4$ means that there exists a positive constant K, depending upon u, such that the difference between u at the $x + \Delta x$ node and the first three terms of the expansion, all evaluated at the xth node, is numerically less than $K(\Delta x)^3$ for all sufficiently small Δx.

Finite Difference Operators

Solving equation 2.12 for $\partial u/\partial x$ and dropping all higher-order terms yields the forward difference operator

$$\frac{\partial u}{\partial x} = \frac{u(x + \Delta x) - u(x)}{\Delta x} + O(\Delta x). \tag{2.14}$$

Similarly, equation 2.13 yields the backwards difference operator

$$\frac{\partial u}{\partial x} = \frac{u(x) - u(x - \Delta x)}{\Delta x} + O(\Delta x). \tag{2.15}$$

And subtracting equation 2.13 from equation 2.12 yields the centered difference operator

$$\frac{\partial u}{\partial x} = \frac{u(x + \Delta x) - u(x - \Delta x)}{2\Delta x} + O(\Delta x^2). \tag{2.16}$$

Note that the forwards and backwards approximations are first-order accurate, whereas the centered is second-order accurate. This means that for the same Δx, it should be more accurate. Forwards and backwards difference schemes of higher-order accuracy are possible, too, but they use values of u at two and three Δx away, making them more computationally expensive and difficult to use near boundaries of the computational domain. Likewise, if equation 2.12 and equation 2.13 are added, the resulting equation can be solved for $\partial^2 u/\partial x^2$ such that

$$\frac{\partial^2 u}{\partial x^2} = \frac{u(x + \Delta x) - 2u(x) + u(x - \Delta x)}{\Delta x^2} \tag{2.17}$$
$$+ O(\Delta x^2).$$

This approximation uses three nodes and is centered in space. There are many more possibilities. For an extensive listing of finite different approximations to various differentials, see Hoffmann and Chiang (2000) and Fletcher (1991).

Explicit Schemes

The next step in formulating a finite difference approximation to the one-dimensional (1-D) diffusion equation is to substitute the above definitions of the derivatives into equation 2.11. Before we do so, it is convenient to change the notation such that $u(x + \Delta x)$ is represented by u_{j+1} and $u(t + \Delta t)$ is represented by u^{n+1} where $t = n\Delta t$ and $x = j\Delta x$ and n and j are the integer series $0, 1, 2, 3, \ldots$. Letting $u = T$, substituting equation 2.14 and equation 2.17 into equation 2.11, and solving for the unknown values of T at the new time step yields the forward-in-time, centered-in-space (FTCS) finite difference scheme

$$T_j^{n+1} = s T_{j-1}^n + (1 - 2s) T_j^n + s T_{j+1}^n, \tag{2.18}$$

where

$$s = D\frac{\Delta t}{\Delta x^2}$$

is called the *diffusion number*.

Inspection of equation 2.18 indicates that values of T known at time n are used to approximate the second derivative of T with respect to x (i.e., its curvature). This quantity added to the value of T_j^n (represented by the coefficient 1 in the second term on the right-hand side) provides an estimate of how T changes from its initial value over one time step. The grid in figure 2.2 can be swept from $j = 2$ to $j = \text{JM} - 1$ for each successive time step. Values of the function at $j = 1$ and $j = \text{JM}$ are not computed because they are known from the boundary conditions (presuming Dirichlet boundary conditions). Notice that the scheme is set up to estimate the value of the function at $f(j, n + 1)$ by using the value of the function at $f(j, n)$, $f(j - 1, n)$, and $f(j + 1, n)$ (i.e., by using values that are all known at the time of the computation). For that reason the scheme is called *explicit*.

One can imagine, however, that the structure of the solution surface as plotted in x, t space may look like a topographic map with domes and hollows. Explicit schemes estimate the curvature of the solution in space *not* at the same time as we want the solution but at an earlier time when the geometry of the solution surface may be different. Wouldn't it be better (more internally consistent) to estimate the curvature at the same point in the space–time plane as the temporal derivative? Usually the answer is yes. We say usually, because the most obvious approach to effect this for equation 2.11 is to use a centered difference operator for the time derivative. Unfortunately, this centered in time and centered in space (CTCS) scheme, also called the Richardson scheme, is unconditionally unstable and of no practical use.

Alternatively, one could estimate the curvature at the point in the space–time plane at the new time step where the new value of T is being computed. This leads to *implicit* schemes as the following example demonstrates.

Implicit Schemes

We rewrite equation 2.18 to estimate the curvature at the $n + 1$ time step. Gathering all unknowns on the left-hand side (LHS):

$$-sT_{j-1}^{n+1} + (1+2s)T_j^{n+1} - sT_{j+1}^{n+1} = T_j^n. \tag{2.19}$$

This is called the Laasonen fully implicit scheme. But now there are three unknowns and only one equation. The path out of this dilemma is to notice that we can write equation 2.19 for each node in space (at the $n + 1$ time step), thereby generating just enough equations for the number of unknowns. Figure 2.3 shows a simple example of two unknown points surrounded by known values provided by the boundary and initial conditions: Writing equation 2.19 at $n = 2$ and at $j = 2$ and then at $j = 3$ yields

$$-sa + (1 + 2s)T_2^2 - sT_3^2 = c$$
$$-sT_2^2 + (1 + 2s)T_3^2 - sf = d, \tag{2.20}$$

where the exponents on T indicate time step 2 (not a squaring operation), which assembled in matrix notation becomes

$$\begin{pmatrix} (1+2s) & -s \\ -s & (1+2s) \end{pmatrix} \begin{pmatrix} T_2^2 \\ T_3^2 \end{pmatrix} = \begin{pmatrix} c + sa \\ d + sf \end{pmatrix}. \tag{2.21}$$

Thus the resulting equations constitute a linear system that can be solved by matrix methods.

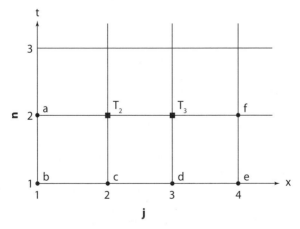

Figure 2.3. Example problem grid for two nodes solved by the fully implicit method.

Another very popular implicit scheme is the Crank–Nicolson scheme. It uses the logic that a forward-in-time approximation of the time derivative is actually estimating the slope of the function with respect to time at a point halfway between n and $n + 1$. Consequently, we should center our estimate of the curvature on that point, too, and that means calculating the curvature at time n *and* at time $n + 1$ and averaging the two. The resulting computation equation is

$$-0.5sT_{j-1}^{n+1} + (1+s)T_j^{n+1} - 0.5sT_{j+1}^{n+1}$$
$$= 0.5sT_{j-1}^n + (1-s)T_j^n + 0.5sT_{j+1}^n. \tag{2.22}$$

Some schemes weight the estimate of curvature a little toward the $n + 1$ time step by using proportions other than 0.5. In any case, equation 2.22 written for each node in the j direction will contain an unknown at $j - 1$, j, and $j + 1$ of the form

$$a_j T_{j-1} + b_j T_j + c_j T_{j+1} = d_j. \tag{2.23}$$

When are all assembled in matrix form, the structure is

$$\begin{bmatrix} b_1 & c_1 & & & 0 \\ a_2 & b_2 & c_2 & & \\ & a_3 & b_3 & \cdot & \\ & & \cdot & \cdot & c_{n-1} \\ 0 & & & & b_n \end{bmatrix} \begin{bmatrix} T_1 \\ T_2 \\ \cdot \\ \cdot \\ T_n \end{bmatrix} = \begin{bmatrix} d_1 \\ d_2 \\ \cdot \\ \cdot \\ d_n \end{bmatrix}. \tag{2.24}$$

Notice that the first or coefficient matrix is tridiagonal; that is, it contains terms only along the center and adjacent two diagonals. Tridiagonal systems of equations like this can be solved efficiently using a simplified form of Gaussian elimination known as Thomas' algorithm. An insightful discussion on this very efficient form of Gaussian elimination and solvers in C and Fortran can be obtained at http://www.nr.com/.

A handy method of summarizing finite difference schemes is to present the basic computational module graphically. Figure 2.4 shows the templates for the four schemes discussed above.

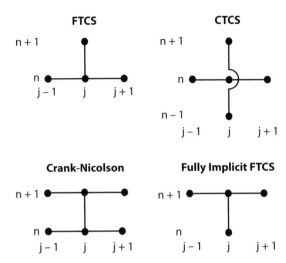

Figure 2.4. Nodes used in approximations to differentials of a one-dimensional diffusion equation. FT, forward in time; CS, centered in space; CT, centered in time.

How Good Is My Finite Difference Scheme?

A choice of finite difference schemes begs the question of which scheme is better. Better can be defined in a number of ways, but we define it as a scheme that is *accurate, efficient,* and *easy.* A scheme is accurate (also called convergent) if its solution approaches the analytic solution as the discretization steps are reduced in size. A scheme is efficient if it minimizes computation time. Easy refers to our ability to comprehend and code the scheme. Judging a scheme's efficiency and ease of use is straightforward, but how do we guarantee its accuracy? The answer is that we must guarantee its *consistency* and *stability.*

A scheme's system of algebraic equations is consistent if, as $\Delta x, \Delta t \to 0$, the system becomes equivalent to the differential equations (DEs) at each grid point. To determine consistency, expand the finite difference equation about x

and t by Taylor series to recover the ODE or PDE. In addition, there will be a remainder of higher-order terms. If the remainder tends to zero as Δx, $\Delta t \to 0$, then the finite difference equation is consistent.

A scheme is stable if spontaneous perturbations (such as round-off error) in the solution of the algebraic equations decay as the computations proceed. A variety of methods exist to determine a scheme's stability such as a von Neumann stability analysis, but these are beyond the scope of this book. See Fletcher (1991) or Hoffmann and Chiang (2000) for excellent summaries.

Finally, a solution of the algebraic equations approximating a DE is convergent if the approximate solution approaches the exact solution as grid size tends to zero. To guarantee convergence, we make use of the Lax equivalence theorem. The Lax equivalence theorem states: *"Given a properly posed linear initial value problem, if a finite difference approximation is consistent and stable, it is convergent."*

Figure 2.5 summarizes these concepts.

So which schemes in figure 2.4 are better? Generally, it can be said that higher-order approximations are more accurate unless they are (1) unstable or (2) the exact solution contains discontinuities or steep gradients. For example, the CTCS (Richardson) scheme would seem to be better than the FTCS explicit scheme because it is of

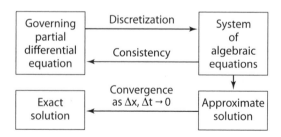

Figure 2.5. Relationship between consistency, stability, and convergence. [Modified from Fletcher, C.A.J. (1991). *Computational Techniques for Fluid Dynamics*. Berlin, Springer-Verlag.]

second order in both time and space discretization (equation 2.16 and equation 2.17), whereas the FTCS scheme is only first-order accurate in time. But the CTCS scheme is unconditionally unstable. The FTCS scheme is stable under certain conditions. As indicated by a von Neumann stability analysis, the explicit FTCS scheme approximating equation 2.11 is stable if

$$s = \frac{D\Delta t}{\Delta x^2} \leq 0.5. \tag{2.25}$$

The fully implicit FTCS and Crank–Nicolson schemes are second-order accurate in both time and space and unconditionally stable.

Stability Is Not Accuracy

As an example of how stability depends upon s, consider solutions to the 1-D diffusion equation describing viscous flow of a Newtonian fluid adjacent to a solid wall. At $t > 0$ the wall at $x = 0$ begins to move instantaneously at a velocity V_0. If the resulting flow is nonturbulent as it is dragged along, then the equation describing the fluid velocity parallel to the wall at various distances y away from the wall, $V(y)$, is described by

$$\frac{\partial V}{\partial t} - \nu\frac{\partial^2 V}{\partial y^2} = 0, \tag{2.26}$$

where ν is the kinematic viscosity of the fluid. This equation is derived in chapter 4. Note the similar form to equation 2.11. There is an analytic solution to this problem given by

$$V = V_0\left\{\sum_{n=0}^{\infty} erfc\,[2n\eta_1 + \eta] - \sum_{n=0}^{\infty} erfc\,[2(n+1)\eta_1 - \eta]\right\}, \tag{2.27}$$

where

$$\eta_1 = \frac{h}{(2\sqrt{\nu t})}$$

$$\eta = \frac{y}{(2\sqrt{\nu t})}.$$

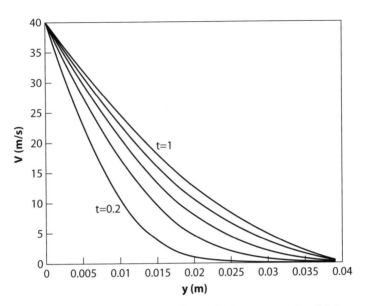

Figure 2.6. Solid lines are analytic solutions to equation 2.26 under the ICs and BCs specified in the text. Dashed lines are solutions from the FTCS scheme with $s = 0.2$. The dashed and solid lines are indistinguishable.

h is the thickness of the fluid, and *erfc* is the complementary error function.

For the purposes of comparing finite difference solutions with the analytic solution, consider the particular problem of an oil of kinematic viscosity equal to 2×10^{-4} m^2 s^{-1} sitting in a 40-mm-thick space bounded by a fixed wall at $y = 0.04$ m and a wall at $y = 0$ that at $t > 0$ begins to move at $V_0 = 40$ m s^{-1}. The analytic solutions are given in figure 2.6 at five equally spaced times from 0.2 to 1 second, showing the development of the velocity profile toward a steady state. Also shown are the numerical solutions to equation 2.26 obtained by the explicitly FTCS scheme with $s = 0.2$. The numerical solution is indistinguishable from the analytic solution.

However, inaccuracies appear as s is increased. The difference between the analytic and numerical solutions

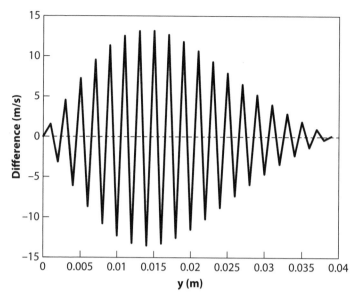

Figure 2.7. Difference between the analytic and FTCS solutions to equation 2.26 for the velocity profile in a viscous fluid. ICs and BCs defined in text. Dashed line computed at $t = 0.5$ second with a diffusion number, $s = 0.5$; solid line computed at same time with $s = 0.508$, illustrating that s must be less than or equal to 0.5 for stability of the FTCS scheme.

to equation 2.26 at $t = 0.5$ and $s = 0.5$ remain small (fig. 2.7). Remember that a von Neumann stability analysis shows that s must be less than or equal to 0.5 for stability of the FTCS scheme. With a slight increase in the diffusion number to 0.508, the scheme becomes unstable and therefore inaccurate.

Summary

A numerical solution to a differential equation or equation set is obtained by converting the equations into an algebraic equation or equation set. This is accomplished by approximating values of the derivatives using Taylor series.

Then the resulting algebraic equations are solved for the dependent variables either explicitly or implicitly at discrete points in the space–time plane. Subsequent chapters will first (and foremost) focus on the process of translating natural phenomena into sets of differential equations but then also explore how numerical solutions to these equations can be obtained. We begin with problems that through translation lead to ordinary differential equations.

Modeling Exercises

1. **Matrix Algebra**
   Consider the set of equations

   $3x + 2y + 5z = 0$
   $7x + 6y + 4z = -2$
   $x + 3y + 2z = -6.$

   Write the set in matrix form $\mathbf{Ax} = \mathbf{b}$. Is the $\mathbf{A}$ matrix invertible? Use your favorite math software (MATLAB, Mathematica, etc.) to solve the equation set for x, y, and z.

2. **The First Numerical Model**
   Write a simple code to calculate the time evolution of the viscous velocity profile that arises from the numerical solution to equation 2.26. Use the FTCS scheme given by equation 2.18, the initial and boundary conditions given in the text, and a diffusion number of 0.2. You will have an outer loop (for loop or do loop) that progresses through time (the n index) and an inner loop that sweeps the grid from left to right (the j index). Your solutions should be identical to those given in figure 2.6. Then reproduce the analysis of figure 2.7.

3. **Practice with Implicit Schemes**
   Now apply the fully implicit scheme (equation 2.22) to the viscous velocity profile problem discussed in the text. Assess its accuracy for various diffusion numbers.

Box Modeling: Unsteady, Uniform Conservation of Mass

We start our discussion of model derivations with systems that are best considered in terms of macroscopic control volumes, or "boxes"; that is, large reservoirs of mass or energy that are effectively homogeneous (well mixed) and evolve in time in response to imbalances between input and output. A familiar example is the global carbon cycle, which one typically envisions as a set of carbon reservoirs, ocean, atmosphere, living organisms, sediments, soils, and sedimentary rocks, among which carbon is transferred by a host of physical and biological processes. In such problems we are generally uninterested in spatial distributions within the reservoirs, although we may wish to study the coupled response of many such reservoirs to internally or externally driven perturbations. In the case of the carbon cycle, we may wish to separate the global ocean into surface, deep, and high-latitude boxes; if we do, however, we must specify the water fluxes that advect carbon from one oceanic reservoir to the next.

Consideration of conservation of mass or energy in systems of reservoirs leads to a set of coupled ordinary differential equations. The process of constructing and solving such systems is often called *box modeling*. In this chapter, we describe the process of box modeling through a number of examples, including an assessment of the controls on the radiocarbon content of the biosphere, a simplified version of the global carbon cycle, a method for interpreting excursions in the carbon isotopic composition of the ocean in deep time, and a simple climate model. Along the way we will introduce the important concepts of residence and response time, steady state, coupled systems, and nonlinear systems. We then present methods for the solution of the ODEs that arise. For further examples and an alternative introduction to box modeling for geochemical cycles, see Walker (1991).

Translations

Example I: Radiocarbon Content of the Biosphere as a One-Box Model

Physical Picture

Cosmic ray bombardment of the atmosphere leads to the production of radioactive ^{14}C from the abundant ^{14}N nucleus. The rate of production thus varies as a function of the cosmic-ray neutron flux to the atmosphere, which varies in time, and the abundance of nitrogen, which is effectively constant over the time scales of interest (millennia). ^{14}C is radioactive and thus is lost from its Earth surface reservoirs through radioactive decay. Radiocarbon's abundance, M, is often characterized in terms of "radiocarbon units" (RCU = 10^{26} ^{14}C atoms). Accordingly, the rate of decay (D) and the rate of production (P) are expressed in units of RCU y^{-1}. Radiocarbon produced in the atmosphere rapidly combines with oxygen to form CO_2 and then gets photosynthesized or stirred into the ocean; most radiocarbon (92%) resides in the deep ocean. We define the *biosphere* (after Vernadsky, 1997 reprint) as the atmosphere, ocean, and living and decomposing biomass. The radiocarbon content of the biosphere, M, can be modeled with a box model (fig. 3.1).

Physical Laws

Radiocarbon decays with a known rate that is linearly proportional to its abundance (fig. 3.2), that is,

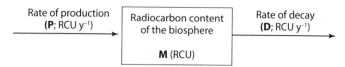

Figure 3.1. Radiocarbon balance for the biosphere; M changes with time in response to imbalances between rate of production, P, and rate of decay, D.

$$D = \frac{dM}{dt} = -kM, \tag{3.1}$$

where the minus sign indicates decay, and k is the decay constant ($k = 1.209 \times 10^{-4}$ y^{-1}). If a sample is isolated from its source of production, for example, when photosynthesis by a cotton plant leads to incorporation of radiocarbon into the cotton, equation 3.1 can be integrated to yield

$$M = M^0 e^{-kt}, \tag{3.2}$$

where M^0 is the initial amount of radiocarbon in the sample (RCU), and t is the time (in years). This equation

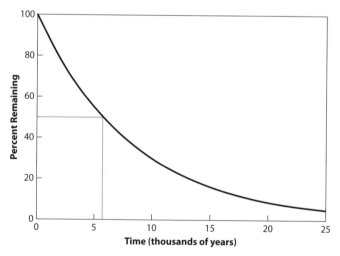

Figure 3.2. Decay of an initial amount of radiocarbon isolated from the atmospheric source. The half-life (the time it takes for half of the material to decay), 5,730 years, is shown.

indicates that the radiocarbon content simply decreases with time according to the well-known exponential decay law. Our ODE and its solution will be a bit more complicated because of the continuous production term P.

Restrictive Assumptions

We are assuming that the biosphere is homogeneous with respect to its radiocarbon content and that all other processes in the carbon cycle are either unimportant or balanced so that we can justifiably focus on the balance between production via cosmic rays and consumption via radioactive decay. For now, also assume that the rate of production is constant in time. Finally, we acknowledge that decay rate constants are constant in time and independent of any other physical condition of the environment.

Perform the Balance

Having defined the processes that introduce or remove radiocarbon from the atmosphere, we can perform the mass balance, first in words, and then symbolically. We write:

$$TROCM = MRI - MRO + SOURCE/SINK.$$

TROCM is shorthand for "the time rate of change of mass in the control volume" (in our case, a macroscopic reservoir of mass), MRI stands for "mass rate into the control volume," and MRO stands for "mass rate out." Sources and sinks reflect any internal production and destruction. In this example, there is no transport in or out of radiocarbon, so we only have internal sources and sinks.

The time rate of change of mass of radiocarbon in the biosphere is equal to the source (production by cosmic rays from nitrogen) minus the sink (radioactive decay):

$$\begin{aligned} \frac{dM}{dt} &= P - D \\ &= P - kM. \end{aligned} \qquad (3.3)$$

Check Units

All terms in the equation above are expressed in units of RCU y^{-1}.

Define Interval, Specify Initial and Boundary Conditions

For time-dependent ODEs (even systems of ODEs), we only need to provide initial conditions, because there are no explicit spatial boundaries to the reservoirs. We do have to provide initial conditions for each reservoir being simulated. In this case, for heuristic purposes, we could consider a biosphere that is suddenly exposed to cosmic rays, with an initial biospheric radiocarbon content of zero, so that we could observe its temporal evolution toward a constant value consistent with modern rates of production and its known decay constant.

A reservoir is said to be in *steady state*, that is, unchanging in time, when its inputs and outputs are balanced:

$$\frac{dM}{dt} = \text{input rate} - \text{output rate} = 0. \tag{3.4}$$

In this case, at steady state:

$$\frac{dM}{dt} = P - kM = 0. \tag{3.5}$$

Rearranging terms, we can solve for the steady-state radiocarbon abundance of the biosphere, M^{ss}:

$$M^{ss} = \frac{P}{k}. \tag{3.6}$$

The abundance of radiocarbon (and any radionuclide with a simple production and decay scheme as shown here) at steady state is thus directly proportional to production rate and inversely proportional to its decay constant.

Often in modeling global cycles of the elements, we use the concept of steady state to estimate terms in the mass balance that are otherwise difficult to measure. For example, in this case we could use the measured abundance of radiocarbon in the biosphere and the known decay constant for radiocarbon to determine its average production rate. So, for a steady-state abundance $M^{ss} = 20{,}300$ RCU and a decay constant of 1.209×10^{-4} y^{-1}, equation 3.6 yields a production rate $P = 2.45$ RCU y^{-1} (Lassey and Enting, 1996).

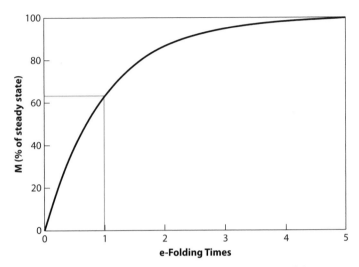

Figure 3.3. The growth of the radiocarbon content of the atmosphere from an initially depleted atmosphere, expressed in e-folding times. For radiocarbon, the e-folding time ($1/k$) is 8,271 years.

There is a known analytic solution to equation 3.5 for a specified initial condition, M^0:

$$M = \frac{P}{k} - \left(\frac{P}{k} - M^0\right)e^{-kt}. \tag{3.7}$$

Note that as

$$t \to \infty, M \to \frac{P}{k} = M^{ss}. \tag{3.8}$$

Thus, the system evolves to the steady state predicted on the basis of a balance between input and output (fig. 3.3). The characteristic time it takes for an initial perturbation from steady state ($M^0 - M^{ss}$) to diminish (i.e., its *response time*) can be characterized by $1/k$ (the inverse of the decay constant, expressed in years). For exponentially decaying reservoirs, the response time is sometimes referred to as the *e-folding time*, because it is the time it takes for the perturbation to diminish by a factor of e^{-1}, or ~37% of its initial value.

For reservoirs at steady state, one can determine the average amount of time a unit of material spends in the reservoir by dividing the reservoir size by either the input or the output (because they are equal at steady state). The result is referred to as the *residence time* (τ). In the case above,

$$\frac{M^{ss}}{P} = \frac{1}{k} \equiv \tau. \tag{3.9}$$

Note here that the residence and response times are the same. This is not generally true, especially for reservoirs that have multiple inputs and outputs. In such cases, the residence time can only be defined with respect to a particular input or output, and the response time is different from any of these residence times.

Periodic Forcing

A more interesting situation arises if the input to the reservoir varies in time (Holland, 1978). For example, the radiocarbon production rate varies periodically in time with the sunspot cycle (11-year period). We can represent this with a production rate function that varies about our canonical value of $P' = 2.45$ RCU y^{-1} with an amplitude $b = 1$ RCU and a frequency

$$\omega = \frac{2\pi}{11} \tag{3.10}$$

as

$$P = P' + b\sin(\omega t), \tag{3.11}$$

where t is time in years. The differential equation is now:

$$\frac{dM}{dt} = P' + b\sin(\omega t) - kM, \tag{3.12}$$

which has an analytic solution (for an initial value of M^0)

$$\begin{aligned} M = \frac{P'}{k} &- \left(\frac{P'}{k} - M^0 - \frac{b\omega}{k^2 + \omega^2}\right)e^{-kt} \\ &+ \frac{b}{\sqrt{k^2 + \omega^2}}(\sin\omega t - \delta). \end{aligned} \tag{3.13}$$

Here the phase lag, δ, between the forcing and the response in M is

$$\delta = \cos^{-1}\left|\frac{k}{\sqrt{k^2 + \omega^2}}\right| \qquad \left(0 \le \delta \le \frac{\pi}{2}\right). \qquad (3.14)$$

The first term on the RHS of equation 3.13 is the previous steady state determined for constant production. The second term represents the exponential decay of any initial deviation from steady state; note that there is a perturbation (source) introduced during this transient period from the presence of the oscillating production term. The third term persists indefinitely and represents oscillation about the previous steady state introduced by the sinusoidal production term. The amplitude of the oscillations is proportional to the product of the amplitude of the production rate variations and the rms time constants (the inverse of the root mean square of the first-order decay constant and the production frequency) of the system. Note that when the frequency of the production variations is small (i.e., when $\omega \to 0$ and $b \sin \omega t$ becomes $<< P'$, equation 3.13 reverts to equation 3.7 because the sinusoidal variations are occurring on a longer time frame that the time frame of interest and thus are undetectable. As Bryan Gregor has noted, "The recognition of a process as secular or cyclic is partly a matter of time. The butterfly may not believe a rabbit who tells her that the seasons recur and that he is looking forward to the next spring" (Gregor, 1988). Perhaps unexpectedly, as the frequency becomes large (i.e., as ω becomes $>> k$), equation 3.13 again reverts to equation 3.7. In this case, the fluctuations are happening too quickly to be detected, as in the *Star Trek* episode "Wink of an Eye," where Captain Kirk is accelerated to the high-frequency Scalosian universe that moves so rapidly that it is only recognizable as a buzzing to the crew of the *Enterprise*.

Note also that the phase lag between changes in production (the forcing) and changes in reservoir size (the response) depend on the relationship between the frequency of the production variations and the decay constant of the reservoir. When $\omega >> k$ (i.e., when the period of the forcing is much shorter than the response time of the reservoir), the reservoir responds with a phase lag of $\pi/2$, or a quarter phase lag of the period of the forcing. This is the maximum lag possible between forcing and response in a simple, linear system (Richter and Turekian, 1993). For sinusoidal forcing, this means that the reservoir responds with its first derivative to the forcing (i.e., it increases most rapidly when the forcing has the largest positive value, and vice versa). In contrast, when the period of the forcing is long with respect to the response time of the reservoir, the phase lag is 0; the reservoir rises and falls in concert with the forcing.

So, in the case of radiocarbon in the biosphere, sunspot cycles (11-year period) have a significant effect on production rates but, once the transient effect of imposing a sinusoidal production term has vanished, we are left with a barely noticeable effect on the radiocarbon abundance (fig. 3.4). The radiocarbon abundance responds as predicted with its first derivative to the forcing.

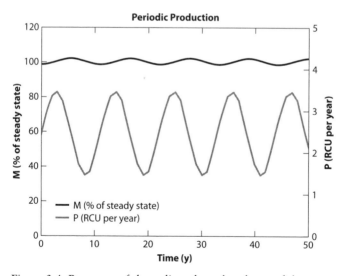

Figure 3.4. Response of the radiocarbon abundance of the atmosphere (left ordinate, expressed as a percentage of the average steady-state abundance) to the 11-year sunspot cycle, presuming an amplitude of 1 RCU to the variation in production rate (right ordinate).

Example II: The Carbon Cycle as a Multibox Model

In the example above, only one reservoir was considered—the radiocarbon content of the atmosphere. More commonly, we are interested in the transfer of matter or energy between two or among multiple reservoirs. In such cases, we end up with a system of coupled ordinary differential equations. Here we explore two example cases, one of matter transfer (the global carbon cycle) and another of energy transfer (a simple 1-D climate model).

Physical Picture

For simplicity, we begin by considering an isolated part of the global carbon cycle, one that involves only two reservoirs, the atmosphere and the living biomass (after Holland, 1978). The atmosphere contains carbon in the form of gaseous carbon dioxide, whereas living biomass contains carbon in multiple organic forms. Let's call the amount of carbon in the atmosphere and the biomass M_1 and M_2, respectively, with units of gigatons (10^{15} g) of carbon. F_{12}, the flux of carbon from reservoir 1 to 2 due to photosynthesis, removes carbon from the atmosphere and incorporates it into biomass. During respiration and decomposition (F_{21}), the carbon is returned to the atmosphere. The rates of these processes can be expressed in gigatons carbon per year (GtC y^{-1}). If we assume homogeneity in these two reservoirs, we can represent this carbon cycle as a box model with two boxes, coupled by two transfers (fig. 3.5).

It is clear from this diagram that the cycle is closed: carbon is recycled between the two reservoirs but is neither added to nor lost from the system (a gross simplification, of course).

Physical Laws

For this simple consideration, let's assume that the rate of removal of carbon from each reservoir is simply proportional to its mass, as in the first example. In other words, the rate of photosynthesis is

$$F_{12} = k_1 M_1, \tag{3.15}$$

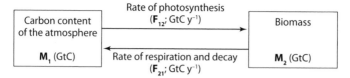

Figure 3.5. Simple representation of the exchange of carbon between the global living biomass and the atmosphere.

and the rate of respiration and decomposition is

$$F_{21} = k_2 M_2. \tag{3.16}$$

Here, k_1 is the rate constant for photosynthesis, and k_2 is the rate constant for respiration and decay, both in units of y^{-1}. We can estimate these rate constants by setting up a steady-state model with specified reservoir sizes and fluxes. Some representative values for today are $M_1 = 800$ GtC, $M_2 = 600$ GtC, and $F_{12} = F_{21} = 60$ GtC y^{-1}. With these values, k_1 becomes 0.075 y^{-1} and k_2 becomes 0.1 y^{-1}. The residence time for reservoir 1 (τ_1) is $1/k_1 = 13.3$ years and for reservoir 2 (τ_2) is $1/k_2 = 10$ years. It's a remarkable fact that the biota process the entire mass of C in the atmosphere in a matter of decades.

Restrictive Assumptions

We are assuming that the atmosphere and biomass are homogeneous in their carbon content and that all other processes in the carbon cycle are either unimportant or balanced so that we can justifiably focus on the balance between photosynthesis and respiration/decomposition.

Perform the Balance

Stated in words, the time rate of change of mass of carbon in the atmosphere is equal to the mass rate in (through respiration and decay) minus the mass rate out (through photosynthesis), or

$$\begin{aligned} \frac{dM_1}{dt} &= F_{21} - F_{12} \\ &= k_{21} M_2 - k_{12} M_1. \end{aligned} \tag{3.17}$$

The time rate of change of mass of carbon in the biomass is simply the negative of equation 3.17, which of course it must be, as we are treating this as a closed system:

$$\frac{dM_2}{dt} = F_{12} - F_{21}$$

$$= k_{12}M_1 - k_{21}M_2. \tag{3.18}$$

Check Units

All terms in the equation above are expressed in units of GtC y^{-1}.

Define Interval, Specify Initial and Boundary Conditions

Here we have two ODEs, so we need to provide initial conditions for M_1 and M_2. Let's assume that at some time (say during an asteroid impact), half of the carbon originally in the biomass is transferred instantaneously to the atmosphere (i.e., $M_1^0 = 1,100$ GtC and $M_2^0 = 300$ GtC).

Under these ICs , the simple, linear, coupled system of equations (equation 3.17 and equation 3.18) has analytic solutions:

$$M_1^t = \frac{k_{21}\left(M_1^0 + M_2^0\right)}{k_{12} + k_{21}} + \frac{k_{12}M_1^0 - k_{21}M_2^0}{k_{12} + k_{21}} e^{-(k_{12} + k_{21})t}$$

and

$$M_2^t = \frac{k_{12}\left(M_1^0 + M_2^0\right)}{k_{12} + k_{21}} - \frac{k_{12}M_1^0 - k_{21}M_2^0}{k_{12} + k_{21}} e^{-(k_{12} + k_{21})t}.$$

Notably, now the response time is $1/(k_1 + k_2)$, which in our case is $1/(0.075 + 0.1) = 5.71$ years (fig. 3.6). Thus, by coupling the two reservoirs, the response time has become shorter than the residence time of either reservoir. This tells us that it is unwise to use residence times as an indicator of reservoir response in even slightly complex box models.

One may argue that it would be more reasonable to specify that the photosynthetic rate depends not only on the atmospheric carbon content but also on the amount of biomass itself. In other words, we might want to express F_{12} as:

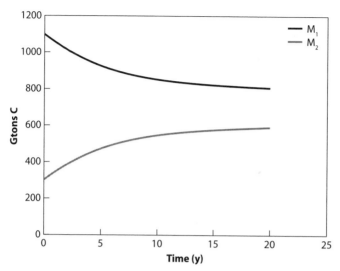

Figure 3.6. Response of the linear C cycle to an initial transfer of 300 GtC from the biomass to the atmosphere. Note that the perturbation from steady state has an e-folding time of 5.7 years, the calculated response time of the system (see text).

$$F_{12} = k_{12}M_1M_2. \tag{3.19}$$

The rates of change of the two reservoirs now become

$$\frac{dM_1}{dt} = k_{21}M_2 - k_{12}M_1M_2 \tag{3.20}$$

and

$$\frac{dM_2}{dt} = k_{12}M_1M_2 - k_{21}M_2. \tag{3.21}$$

Note that k_{12} no longer has intuitive units (GtC^{-1} y^{-1}). Finding analytic solutions for nonlinear systems of equations is difficult, and often impossible. In the next chapter, we will explore numerical solutions to nonlinear ordinary differential equations. Here we simply present the results of the numerical integration (fig. 3.7). Note the interesting difference between this result and the previous simulation of the linear system. Now the response time is considerably

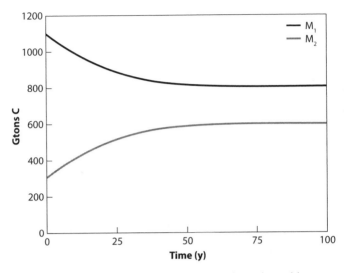

Figure 3.7. Response of the nonlinear C cycle to the sudden transfer of 300 GtC from the biomass to the atmosphere. Contrast with response of linear system in figure 3.6.

longer (~23 years) than that calculated for the linear system (5.7 years) or for either reservoir's residence time (10–13 years). This extended response time arises because one of the two fluxes is sensitive to both reservoirs. Whereas in the linear model the flux from the atmosphere to the biomass increased to more than 80 GtC y^{-1} immediately after the transfer of 300 GtC to the atmosphere, here it *decreased* from the steady-state value of 60 GtC y^{-1} to just slightly over 40 GtC y^{-1}, much closer to the rate of transfer of C from the biomass to the atmosphere (30 GtC y^{-1}), which has been reduced to half the original steady-state value with a halving of the biomass reservoir. The reservoirs evolve back to steady state more slowly then, responding to this much smaller flux imbalance between input and output for each reservoir. In general, if the fluxes are dependent on both reservoir sizes in a coupled system such as this, and if the residence times of the two reservoirs is vastly different, the response time can greatly exceed either of the two residence times (cf. Rothman et al., 2003).

Example III: One-Dimensional Energy Balance Climate Model

Most serious climate modeling is done on supercomputers that solve large systems of partial differential equations simultaneously. However, "toy" climate models, the sort that can easily run on your PC, are sometimes of use in exploring the basics of climate system operation. One version of a toy climate model treats Earth's surface energy as being enclosed in a series of reservoirs that encircle the earth but extend a finite distance in latitude. These "zonal" reservoirs are linked by the exchange of energy between adjacent reservoirs. Although we said above that transport modeling is typically not appropriately treated with box models, in cases like this, where we can specify that the transport of material or energy is dependent on reservoir size, box modeling can be performed. In this example, we draw heavily on the model description by Walker (1991).

Physical Picture

In our one-dimensional energy balance climate model, we separate the earth surface into n zonal bands ($180/n$) degrees wide spanning from pole to pole (fig. 3.8). Each of these reservoirs contains heat energy, characterized by its temperature. When we go to solve this problem, we will need to remember that the volume of each reservoir diminishes with the cosine of latitude, as do the boundaries between adjacent reservoirs across which they exchange energy. Energy is received from the sun at all latitudes, and a fraction (the *albedo, a*) is reflected back to space. Each reservoir radiates heat to space, and the efficiency of this transfer depends on atmospheric composition (i.e., the greenhouse effect). Temperature differences between adjacent reservoirs drives heat transfer (generally from equator to pole).

Physical Laws

Energy input into each box j ($F_{in,j}$), typically in units of W m^{-2}, is calculated as the product of the solar constant (S) and the fraction of this energy absorbed (not reflected, i.e., $1 - a$):

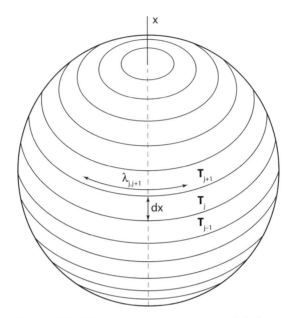

Figure 3.8. Gridded representation of the global climate system. We divide Earth into latitudinal zones of width dx and perform an energy balance for each macroscopic control volume. Note that the circumference of the boundaries between adjacent zones (λ) diminish with the cosine of latitude.

$$F_{in} = S \times (1 - a). \tag{3.22}$$

Note that both S and a are functions of latitude because of variations in average solar angle and differences in land and cloud cover. With simple climate models such as presented here, these values are typically constants obtained from the literature. Outgoing radiation is calculated according to the Stefan–Boltzmann law of blackbody radiation; that is,

$$F_{out} = \varepsilon \sigma T_j^4, \tag{3.23}$$

where σ is the Stefan–Boltzmann constant (5.67×10^{-8} W m^{-2} K^{-4}), and ε is the emissivity (i.e., the efficiency with which the surface is able to emit radiation to space). The emissivity diminishes as the greenhouse effect intensifies (i.e., as CO_2 and other greenhouse gases accumulate in the atmosphere).

The transfer of heat energy from one reservoir to the next occurs as a result of complicated processes of atmospheric circulation. For our purpose, we will assume that the first-order rate law called Fourier's law applies here; that is, that the heat transfer is proportional to the temperature difference between adjacent reservoirs and inversely proportional to the distance between the (centers of) the two reservoirs:

$$F_{x,j,j-1} = -D \frac{T_j - T_{j-1}}{dx} \rho C_p. \tag{3.24}$$

Here, D is the proportionality constant (in units of $m^2\ s^{-1}$), which, when multiplied by the density (ρ; $kg\ m^{-3}$) and heat capacity (C_p; $J\ K^{-1}\ kg^{-1}$) relates the energy flux (in $W\ m^{-2}$) to the temperature gradient. The distance between the centers of the boxes is dx, and T is temperature in Kelvin (K). One typically calculates the heat capacity of each zonal reservoir based on the relative proportions of land and sea in that zone and their different heat capacities.

Restrictive Assumptions

This simple climate model ignores a host of processes that affect Earth's climate. In keeping the albedo constant in each box we ignore any and all feedbacks associated with changing cloud, ice, and vegetation cover. The heat transfer relationship we adopted basically ignores all transfers associated with the general circulation of the atmosphere and with the advection of latent heat (water vapor) by the winds. We also ignore the important effects of topography and other factors that vary longitudinally. The results we obtain can only be interpreted in terms of the model's very basic representation of meridional gradients in temperature.

Perform the Balance

The time rate of change of energy in each reservoir is the difference in the rates of input and output of energy from each reservoir. The energy content of each reservoir is the product of its temperature, density, heat capacity, and volume. Volume is obtained by the product of the surface area of each reservoir j (A_j) and a "radiatively active" thickness H, generally thought of as a depth in the ocean that exchanges heat on a seasonal timescale. The input flux of energy from the sun is through the reservoir area A_j, as is the outgoing infrared radiation F_{out}, but the exchange of energy across latitudinal bands (i.e., from box to box) is through a cross-sectional area that is the product of the depth H and the length of the border between the two adjacent reservoirs, a function of latitude ($\lambda_{j,j-1}$). In other words,

$$\frac{d}{dt}(C_p \rho H A_j T_j) = (F_{\text{in}} - F_{\text{out}}) A_j + F_{x,j,j-1} \\ + F_{x,j,j+1} H \lambda_{j,j+1}.$$ (3.25)

Substituting the expressions for the fluxes (equation 3.22 to equation 3.24) into equation 3.25 gives:

$$\frac{d}{dt}(C_p \rho H A_j T_j) = \left(S \times (1-a) - \varepsilon \sigma T_j^4\right) A_j \\ -D \frac{T_j - T_{j-1}}{dx} \rho C_p H \lambda_{j,i-1} - D \frac{T_j - T_{j+1}}{dx} \rho C_p H \lambda_{j,j+1}.$$ (3.26)

This equation can be simplified by dividing through by the factors that are independent of time (C_p, ρ, H, and A_j) to yield

$$\frac{d}{dt}(T_j) = \frac{\left(S \times (1-a) - \varepsilon \sigma T_j^4\right)}{\rho C_p H} - D \frac{T_j - T_{j-1}}{dx} \frac{\lambda_{j,i-1}}{A_j} \\ -D \frac{T_j - T_{j+1}}{dx} \frac{\lambda_{j,j+1}}{A_j}.$$ (3.27)

Check Units

All terms in the equation above are expressed in units of K y^{-1}.

Define Interval, Specify Initial and Boundary Conditions

A typical initial condition would be to specify a uniform temperature for all reservoirs. Temperatures would then evolve ("spin up") toward the steady state as the result of differential energy input to the various latitudes. Two boundary conditions must be satisfied. Usually, one could either specify the temperature for the two most poleward boxes or, recognizing that there is no flux poleward of these boxes, remove the second term on the RHS of equation 3.27 for reservoir 1 and the first term on the RHS of equation 3.27 for reservoir n.

Once these steps have been completed, we have a well-posed 1-D climate model. The system of equations is highly coupled, nonlinear, and thus not amenable to analytic solution. Instead we turn to numerical solutions.

Finite Difference Solutions of Box Models

As we've seen in past examples, interesting and useful models of Earth systems tend to be nonlinear and highly coupled. To study these models quantitatively usually requires numerical solutions of the system of differential equations. In box models, we have a system of ordinary differential equations to solve beginning from a specified set of initial conditions. As discussed in chapter 2, the general approach is to convert these differential equations to algebraic equations using *finite differences* and then use techniques from linear algebra to solve the equations.

The Forward Euler Method

Conceptually, this approach uses the known derivative of the dependent variable (e.g., y) evaluated at an initial value (e.g., y^n) to extrapolate over a finite increment of the independent variable (e.g., t). In other words,

$$y^{n+1} = y^n + \frac{dy}{dt}\bigg|_{y^n} \Delta t, \tag{3.28}$$

which can readily be generalized in vector notation for systems of equations representing the coupled evolution of multiple reservoirs:

$$\frac{\Delta \vec{y}}{\Delta t} = F(\vec{y}^n, t^n) \tag{3.29}$$

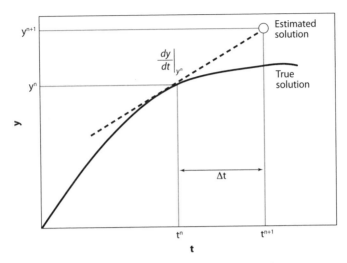

Figure 3.9. The forward Euler method of numerical approximation extrapolates the solution from a known position y^n over a given interval t using the known derivative of the solution (i.e., the tangent to the solution) at that position.

$$\text{where } \Delta\vec{y} = \vec{y}^{n+1} - \vec{y}^n \text{ and } F' = \frac{d\vec{y}^n}{dt}. \tag{3.30}$$

This approach, sometimes referred to as the *forward Euler method*, as well as its shortcomings, are obvious from figure 3.9.

Here it is clear that by picking a large time increment for the extrapolation, we have introduced a large error into our estimate of y at the future time. This error arises because we have implicitly neglected the higher-order terms in the Taylor series approximation used in equation 3.28. If we use the forward Euler approach, we must use very small time steps to avoid large errors. The error is $O(\Delta t)$, so reducing the size of Δt only provides a linear improvement in accuracy.

Notice that if we ignore insolation and infrared radiation, equation 3.27 is similar in form to the FTCS discretization of the one-dimensional diffusion equation (equation 2.18). In fact, if we ignore the variation of λ with latitude, then $A/\lambda = dx$, and we can recast the simplified version of equation 3.27 (diffusion only) as:

$$\frac{d}{dt}(T_j) = -D\frac{T_j - T_{j-1}}{dx^2} - D\frac{T_j - T_{j+1}}{dx^2}$$
$$= D\frac{T_{j-1} - 2T_j + T_{j+1}}{dx^2}. \tag{3.31}$$

Then if we apply the forward Euler method to the solution of equation 3.31, we obtain:

$$T_j^{n+1} = T_j^n + \frac{D\Delta t}{\Delta x^2}\left(T_{j-1}^n - 2T_j^n + T_{j+1}^n\right). \tag{3.32}$$

Equation 3.32 is indeed identical to equation 2.18. In other words, the FTCS solution scheme for the partial differential equation representing one-dimensional diffusion is the same as the forward Euler method of solving the same problem using box modeling. The difference is that in the former we treat Δx as a very small increment in x, whereas in box modeling we consider Δx to be a macroscopic property of the system.

Predictor–Corrector Methods

We could improve on the forward Euler method if we had a way of "anticipating" the curvature of the real solution (i.e., if we could estimate the derivative of the function at a future time). If so, then we could average these two derivatives to obtain a better estimate of the slope of the solution in the interval of extrapolation; that is to say, we want

$$y^{n+1} = y^n + \left(\frac{\left.\frac{dy}{dt}\right|_{y^{n+1}} + \left.\frac{dy}{dt}\right|_{y^n}}{2}\right)\Delta t. \tag{3.33}$$

Of course, we don't know $\frac{dy}{dt}\big|_{t^{n+1},y^{n+1}}$, so let's first "predict" y^{n+1} using Euler's method, and then correct the value by using equation 3.33. The exponential decay equation provides a simple example:

$$\frac{dy}{dt} = -\lambda y. \tag{3.34}$$

From equation 3.28,

$$y^{n+1} = y^n - \lambda y^n \Delta t$$
$$= y^n(1 - \lambda \Delta t). \tag{3.35}$$

This is the Euler predictor step for y^{n+1}. In equation 3.33, call the derivatives two new terms, k_1 and k_2:

$$k_1 = \left.\frac{dy}{dt}\right|_{y^n} \tag{3.36}$$

and

$$k_2 = \left.\frac{dy}{dt}\right|_{y^{n+1}} \tag{3.37}$$

such that

$$y^{n+1} = y^n + \left(\frac{k_1 + k_2}{2}\right)\Delta t. \tag{3.38}$$

For example, if $\Delta t = 1$, $\lambda = 0.25$, and $y^0 = 16$, then for the predictor step, using equation 3.34, $y^1 = 16\,(1 - 0.25 \times 1) = 12$. We use this value to calculate $k_2 = -0.25 \times 12 = -3$; $k_1 = -0.25 \times 16 = -4$. Then for the corrector step we use equation 3.38 to improve on our previous estimate of y^1:

$$y^1 = 16 + \left(\frac{-4 + (-3)}{2}\right)1 = 12.5. \tag{3.39}$$

This method provides an improved estimate of the actual solution; in fact, it coincidentally *is* the actual solution after the first time step. The method is called *second-order Runge–Kutta* after Carl Runge (1856–1927), a German mathematician and astronomer who has a crater on the moon named after him, and another German mathematician, M. W. Kutta (1867–1944), a pioneer in the field of aerodynamics. At subsequent times the approximation is not perfect, but the accuracy is improved over the Euler method (table 3.1); it is now $O(\Delta t^2)$.

Higher-order Runge–Kutta methods add terms (k_3, k_4) that improve the accuracy even more but have additional computational overhead.

Stiff Systems

We have seen that in developing box models for natural systems (e.g., the carbon cycle), we encountered reservoirs

Table 3.1. Comparison of the Numerical Approximations of the Exponential Decay Equation Using Forward Euler and Second-Order Runge–Kutta Methods

t	y (true)	y (forward Euler)	y (second-order Runge–Kutta)
0	16.00	16.00	16.00
1	12.50	12.00	12.50
2	9.70	9.00	9.76
3	7.56	6.75	7.58

coupled by exchange of material that had vastly different time constants (response or residence times). Such instances revealed interesting behaviors in terms of generating much longer period responses to forcings than one would anticipate based on residence times alone.

Mathematicians refer to systems of equations with wide-ranging time constants as *stiff systems*. These systems present particular challenges to numerical solution using the "explicit" methods discussed so far, because stability constraints require that we keep our time increment Δt small with respect to the residence time of the fastest cycling reservoir. In other words, even if we are primarily interested only in the long-term behavior of the system, we must take short time steps (i.e., perform many more calculations than would seem to be necessary to study the system).

One example of a stiff system is the set of equations describing what has been called the "Rothman ocean" (Rothman et al., 2003).

Example IV: Rothman Ocean

Physical Picture

Consider the following model of the oceanic carbon cycle, including two reservoirs: one, the dissolved inorganic carbon reservoir (DIC; M_1), and the second, a large reservoir of dissolved organic carbon (DOC; M_2) thought to have characterized the Proterozoic Era ocean more than

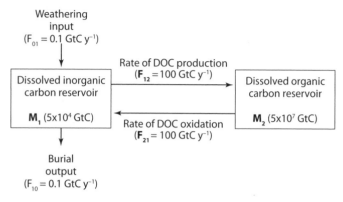

Figure 3.10. Representation of a simplified steady-state, Proterozoic Era "Rothman ocean" (Rothman et al., 2003) with a large reservoir of dissolved organic carbon (M_2) and a small reservoir of dissolved inorganic carbon (M_1), the opposite of today's.

500 million years ago (today the DOC reservoir is much smaller than the DIC reservoir) (fig. 3.10). The ocean is provided a steady supply of carbon from weathering of C-containing rocks and from volcanic eruption, and this is removed by the burial of carbon in sediments (here represented only by inorganic C burial; we neglect organic carbon burial for simplicity here). Dissolved organic carbon is produced by photosynthesis followed by incomplete decomposition; further decomposition (DOC oxidation) regenerates DIC.

Physical Laws

Presuming that the fluxes are all linear with respect to the size of the reservoir from which they emanate, we can use the steady state from figure 3.10 to calculate rate constants and define the following rate relationships: $F_{10} = 2 \times 10^{-6} M_1$, $F_{12} = 2 \times 10^{-3} M_1$, and $F_{21} = 2 \times 10^{-6} M_2$.

Restrictive Assumptions

We are assuming that all carbon is removed from the inorganic C reservoir (presumably buried as $CaCO_3$); a more complete treatment would also consider the burial of organic matter. We are also neglecting the multitude of factors other than reservoir size that control these fluxes.

Perform the Balance

The time rate of change of carbon in the DIC and DOC reservoirs is the mass rate in minus the mass rate out; that is,

$$\frac{dM_1}{dt} = 0.1 + 2 \times 10^{-6}M_2 - 2 \times 10^{-6}M_1 \\ - 2 \times 10^{-3}M_1 \tag{3.40}$$

and

$$\frac{dM_2}{dt} = 2 \times 10^{-3}M_1 - 2 \times 10^{-6}M_2. \tag{3.41}$$

Check Units

All terms in the equation above are expressed in units of GtC y^{-1}.

Define Interval, Specify Initial and Boundary Conditions

We can solve for the steady state of the system by setting equation 3.40 and equation 3.41 to zero, giving us two equations and two unknowns. The resulting steady states are

$$M_1^{ss} = 5 \times 10^4 \text{ GtC}$$

$$M_2^{ss} = 5 \times 10^7 \text{ GtC}.$$

The DOC reservoir is 1,000 times larger than the DIC reservoir at steady state. Of course, we already knew the steady-state values because we used them to calculate the rate constants.

Let us presume we are interested in the response of the oceanic carbon cycle to a doubling of the riverine input F_{10} (from 0.1 to 0.2 GtC y^{-1}). We use the forward Euler method with a time step (Δt) of 100 years (fig. 3.11).

In 10,000 years, reservoir 1 (the DIC reservoir) has adjusted to a new, apparent steady state, whereas the large

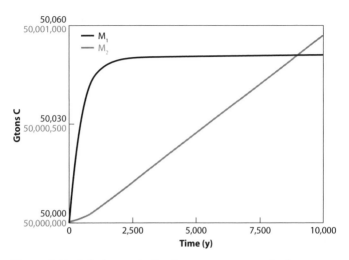

Figure 3.11. Solution to the Rothman ocean perturbation (doubling of river C input) over the first 10,000 years. Note that the DIC reservoir (M_1) has reached quasi-steady-state. Forward Euler method, 100-year time step.

DOC reservoir (reservoir 2) is increasing. We know that the new steady-state sizes of M_1 and M_2 are going to be twice the initial sizes. The doubling of M_2 will take some several hundred million years, though, so we need to perform a longer simulation. However, to reduce the number of computations, we increase the step size (Δt) to 1,000 years and compare this result for the first 10,000 with that shown in figure 3.11. The result is shown in figure 3.12.

No, we haven't discovered a natural behavior worthy of further investigation. We simply revealed numerical instability. This behavior is referred to as "sawtoothing" for obvious reasons. The solution has become unstable, oscillating about the exact solution, with the estimate of the derivative changing signs back and forth introducing large inaccuracies. The source of this problem is more easily seen in the simple case of exponential decay, where the Euler solutions scheme is

$$y^{n+1} = y^n(1 - \lambda\Delta t). \tag{3.42}$$

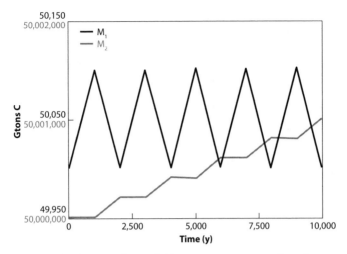

Figure 3.12. Same as figure 3.11, except now a step size of 1,000 years was specified, revealing "sawtoothing" behavior.

Note here that if $\Delta t > 2/\lambda$, the estimate of y changes sign every time step. Increasing the time step even further leads to even more erratic behavior with huge errors. If we wish to use this method, then, we are forced to use small, century-long time increments. But we want to integrate over 100 million years, thereby requiring a million time steps. Is there a more efficient way to perform this calculation that allows larger time steps without jeopardizing stability?

Backward Euler Method

What if we could evaluate the derivative of the function at a future time step? Would knowing that derivative and using it to extrapolate forward in time improve the stability of the solution? To do so, we rewrite equation 3.28 as

$$y^{n+1} = y^n + \left.\frac{dy}{dt}\right|_{y^{n+1}} \Delta t. \qquad (3.43)$$

For the problem of exponential decay (equation 3.34), this becomes

$$y^{n+1} = y^n - \lambda y^{n+1} \Delta t$$
$$= \frac{y^n}{(1 + \lambda \Delta t)}. \tag{3.44}$$

Comparing equation 3.44 with equation 3.45, we note that now there is no apparent instability, and as $\Delta t \to \infty$, the approximation converges on the true solution. Thus this method, known as the *backward Euler method*, is stable and gives accurate solutions at long times (although not on short times, but that is okay because we don't care about the short timescales in this problem).

Of course, we don't know the value of the derivative in the future a priori; equation 3.43, and the backward Euler method it represents, are implicit. However, we can approximate it using Taylor series,

$$\left. \frac{dy}{dt} \right|_{y^{n+1}} \approx \left. \frac{dy}{dt} \right|_{y^n} + \frac{\partial}{\partial y} \left(\left. \frac{dy}{dt} \right|_{y^n} \right) \Delta y, \tag{3.45}$$

ignoring the higher-order terms in the series, and where $\Delta y = y^{n+1} - y^n$. Substituting equation 3.45 into equation 3.43 yields

$$\Delta y = \left[\left. \frac{dy}{dt} \right|_{y^n} + \frac{\partial}{\partial y} \left(\left. \frac{dy}{dt} \right|_{y^n} \right) \Delta y \right] \Delta t. \tag{3.46}$$

Now, gather terms to solve for Δy, because Δy gives us the increment in y that we need to calculate y^{n+1}. After rearranging,

$$\Delta y \left(\frac{1}{\Delta t} - \frac{\partial}{\partial y} \left(\left. \frac{dy}{dt} \right|_{y^n} \right) \right) = \left. \frac{dy}{dt} \right|_{y^n},$$

or

$$\Delta y = \frac{\left. \frac{dy}{dt} \right|_{y^n}}{\frac{1}{\Delta t} - \frac{\partial}{\partial y} \left(\left. \frac{dy}{dt} \right|_{y^n} \right)} \tag{3.47}$$

This is the backward Euler method for a single ODE. We can generalize this for a system of equations by defining the Jacobian matrix J as

$$J \equiv \frac{\partial F'(\vec{y}^n)}{\partial y}$$

and recognizing that I is the identity matrix:

$$\left(\frac{1}{\Delta t}I - J\right)\Delta\vec{y} = F'(\vec{y}^n). \tag{3.48}$$

Note that equation 3.48 is similar to the matrix form of the forward Euler method, except for the presence of the Jacobian matrix. Thus, we can imagine intermediate solution schemes between the forward and backward Euler methods and implement these in our codes by incorporating a parameter θ:

$$\left(\frac{1}{\Delta t}I - \theta J\right)\Delta\vec{y} = F'(\vec{y}^n). \tag{3.49}$$

When $\theta = 0$, we have the forward Euler method; when $\theta = 1$, we have the backward Euler method; and when $\theta = 0.5$, we have the Crank–Nicolson method. Crank–Nicolson blends the two methods and in doing so achieves a higher-order accuracy than either method without seriously compromising the stability attributes of the backward Euler method.

Returning to the Rothman ocean example, the F' vector is

$$F' = \begin{bmatrix} 0.1 + 2\times10^{-6}M_2 - 2\times10^{-6}M_1 - 2\times10^{-3}M_1 \\ 2\times10^{-3}M_1 - 2\times10^{-6}M_1 \end{bmatrix}, \tag{3.50}$$

and the Jacobian matrix is

$$J = \begin{bmatrix} -2\times10^{-6} - 2\times10^{-3} & 2\times10^{-6} \\ 2\times10^{-3} & -2\times10^{-6} \end{bmatrix}, \tag{3.51}$$

so equation 3.49 becomes:

$$\left(\frac{1}{\Delta t}I - \theta\begin{bmatrix} -2\times10^{-6} - 2\times10^{-3} & 2\times10^{-6} \\ 2\times10^{-3} & -2\times10^{-6} \end{bmatrix}\right)\Delta\vec{y}$$
$$= \begin{bmatrix} 0.1 + 2\times10^{-6}M_2 - 2\times10^{-6}M_1 - 2\times10^{-3}M_1 \\ 2\times10^{-3}M_1 - 2\times10^{-6}M_1 \end{bmatrix}. \tag{3.52}$$

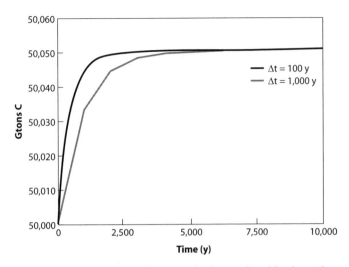

Figure 3.13. Comparison between the forward and backward Euler solutions to the Rothman ocean perturbation over the first 10,000 years, with 100-year and 1,000-year time steps, respectively, for reservoir 1 (the DIC reservoir). Compare with figure 3.11.

This algebraic equation is readily solved using MATLAB or any other linear algebra software that can do matrix inversion. Figure 3.13 compares the short-term (from a geologist's perspective, over 10,000 years) response of the DIC reservoir of the Rothman ocean to a doubling of the riverine input, computed with the forward Euler method (100-year time step) and the backward Euler method (1,000-year time step). Recall that the forward Euler approach exhibited sawtoothing with a time step of 1,000 years (fig. 3.12). The backward Euler method does not sawtooth, but it consistently underestimates the true solution (closely reflected in the forward Euler solution) over the first several thousand years.

However, by the time the DIC reservoir has reached quasi-steady-state (i.e., a slowly evolving steady state), the two solutions have converged. The great advantage of the implicit method (backward Euler) is shown in figure

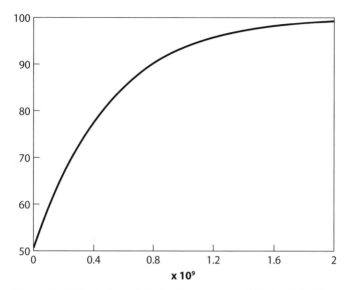

Figure 3.14. Evolution of the Rothman ocean DIC ($\times 10^3$ GtC) and DOC ($\times 10^6$ GtC) in response to a doubling of the riverine C input, calculated using a time step of 100,000 years and the backwards Euler method. The solution is essentially identical with a time step of 1 million years.

3.14, which displays the evolution of the DIC and DOC reservoirs over 2 billion years. This same result is obtained with a time step of 1 million years, reflecting the method's stability and convergence on the true solution on long times.

Model Enhancements

A computer code that incorporates the forward and backward Euler methods is quite flexible and can solve most problems of interest in box modeling. The methods as presented do have some limitations, though. For example, the step size must be specified, and the Jacobian matrix must be obtained analytically. Of course, these problems can be overcome.

Automatic Step-Size Adjustment

For stiff problems, in particular, it may be advisable to adjust the time step as the simulation proceeds, keeping it short at first and then longer as the fast-response reservoirs come to quasi-steady-state. Incorporating automatic step-size adjustment is quite simple. One simply assesses the change in reservoir size from one time step to the next relative to the initial reservoir size (φ):

$$\varphi = \frac{|y^{n+1} - y^n|}{|y^n|}. \tag{3.53}$$

If φ is less than a tolerance limit (ε, often set at 0.001), the step size is increased by an amplification factor, and the next time step is computed. If φ is greater than ε, the step size is diminished by a reduction factor and the time-step recomputed. One can also specify a maximum and minimum value of Δt to prevent round-off and truncation error accumulation. The logic here is that we are trying to keep the change in the solution from one time step to the next small relative to the magnitude of the solution to avoid large errors in the numerical approximation. There are no particular values of the amplification and retardation factors that work for every application, but 0.5 and 2 (halving and doubling) are commonly used.

Numerical Approximation to the Jacobian Matrix
(After Walker, 1991)

Finally, it is sometimes advantageous to make a numerical approximation to the Jacobian matrix when the ordinary differential equations are complex and nonlinear (i.e., when it is difficult or impossible to find the partial derivatives). To do so, first calculate the derivative vector F' for the current values of $\vec{y}_1^n$. Then increment $\vec{y}_1^n$ by a small amount, typically a small multiplier of $\vec{y}_1^n$ (e.g., $\varepsilon^{\vec{y}_1^n}$, where ε is typically 0.001), and recalculate the derivatives. Divide these numbers by $\varepsilon - y_i^n$, change their sign, and to the diagonal term add $1/\Delta t$. This gives you the first column in the Jacobian. Restore the value of $\vec{y}_1^n$ and then repeat the procedure for $\vec{y}_2^n$, and so forth, to fill the Jacobian.

Summary

In this chapter, we explored the translation of simple geologic problems of conservation of mass or energy into systems of ordinary differential equations (i.e., box models). Such models are often called "toy" models, because they are based on significant simplifications of the natural world. However, because they are based on the conservation equations, the results they produce are consistent with these laws, more than can often be said for the "arm waving" conclusions one might make in interpreting data. Thus, they have real value for the geoscientist interested in the big picture and/or in advance of more sophisticated modeling. One shortcoming of box modeling is that it generally specifies, rather than calculates, the physics of flows. For that we turn to spatially dependent systems of equations, partial differential equations, and their solutions: the topics of the remainder of this book.

Modeling Exercises

1. **Runge–Kutta Scheme**
   Write a simple code to solve the problem of radioactive decay numerically using a second-order Runge–Kutta solver (e.g., equation 3.38). Graph the analytic and numerical solutions using the initial condition and decay constant that generated table 3.1. Extra Credit: Find and code a fourth-order Runge–Kutta scheme, and compare its accuracy with that of the second-order scheme here.

2. **The Oceanic Phosphate (P) Cycle**
   a. Construct a simple box model (single ODE) of the P cycle. Follow the proper steps of model building. Use the following information:
      - Average concentration of P in ocean $[PO_4]$ = 2.1×10^{-6} mol kg^{-1} seawater.
      - Mass of the ocean = 14×10^{20} kg.

- Average residence time of P in ocean (with respect to river input or sediment output) = 40,000 years.
- Assume that the sediment output is proportional to the average concentration.

b. Human activity has approximately doubled the input of phosphate to the ocean. Plot the response of the ocean's P content to a doubling of the rate of riverine input. Be sure to carry the calculation out far enough to show the new steady state.

c. Now create a more sophisticated two-box model of the ocean's P cycle. Divide the ocean into a surface box and a deep box. The surface box will have 2.5% of the mass of the deep box and a steady-state concentration 10% of the deep box. The total phosphate content of the ocean (i.e., the average concentration of phosphate) will be identical to that of part 1a, as will the total mass of the ocean. River input will be to the surface box. Use the following additional information to complete the model:

- The removal of phosphate from the surface box with biological productivity is proportional to the concentration of phosphate in that box.
- Ninety-nine percent of this flux is released to the deep ocean during decomposition, and 1% is removed to the sediments (balancing the river input at steady state).
- Water is mixed (upwelled and downwelled) between the surface and deep boxes at a rate of 14×10^{17} kg y^{-1}, creating a net transfer of phosphate from deep to surface that is proportional to this rate, and the concentration difference between deep and surface—that is, the rate (mol y^{-1}) of phosphate added to the surface box—is $(14 \times 10^{17}$ kg y$^{-1}) \times ([PO_4]_d - [PO_4]_s)$.

d. Finally, plot the response of the ocean's phosphate concentration (surface and deep) to a doubling of the mixing rate between surface and deep. Be sure to carry the calculation out far enough to show the new steady state.

3. Stiff Systems and Implicit Solvers

Use the forward and backward Euler scheme to integrate the following set of ODEs (attributed to C. W. Gear) for $0 \leq t \leq 1$ for an initial condition of $y_1(0) = 1$, $y_2(0) = 0$, and for time steps (Δt) of 0.001, 0.0018, 0.002, and 0.1, and compare the results with the analytic solutions:

$$\frac{dy_1}{dt} = 998y_1 + 1998y_2$$

$$\frac{dy_2}{dt} = -998y_1 - 1999y_2.$$

The analytical solutions are

$$y_1 = 2e^{-t} - e^{-1,000t}$$

$$y_2 = -e^{-t} + e^{-1,000t}.$$

4

One-Dimensional Diffusion Problems

There is a very large class of problems in the earth sciences in which a conservative property moves through space at a rate proportional to some gradient (i.e., it follows a first-order rate law). Here, "first order" refers to an equation that contains the first derivative but no higher derivatives. This is different from the use in chemistry where it denotes a reaction rate proportional to the first power of a concentration. The conservative property that flows can be the moles of ions in a solution, or the thermal energy of atoms in a material, or the mass of regolith on a hillside (table 4.1). In the study of transport phenomena, the amount of the property in question that flows per unit area per unit time is called a *flux* (q). If the flux is by number of objects (e.g., moles), the units will be $mol\ m^{-2}\ s^{-1}$; if by volume, $m\ s^{-1}$; and if by mass, $kg\ m^{-2}\ s^{-1}$. In the first case, the quantity of ions would move in proportion to the concentration gradient of those ions; in the second case, the gradient is in temperature; and in the third case, the gradient is the slope of the hill. Regardless of the property in question or the gradient it flows down, all problems of this class acquire the same mathematical form and can be solved in a similar manner. In fact, just knowing that a phenomenon falls in this class allows us to know a lot about its behavior under various initial and boundary conditions, even before solution.

Table 4.1. Typical Flux Laws in Earth Science

1. Mass flux: the rate of mass flow through a unit area orthogonal to the flow [kg m^{-2} s^{-1}]; for example, regolith creep on a hillslope.
2. Momentum flux: the rate of transfer of momentum across a unit area [N m^{-2}]; for example, Newton's law of viscosity.
3. Energy flux: the rate of transfer of energy through a unit area [J m^{-2} s^{-1}]; for example, the rate of heat flow across a unit area as given by Fourier's law of conduction.
4. Chemical flux: the rate of movement of molecules across a unit area [mol m^{-2} s^{-1}]; for example, Fick's law of diffusion.
5. Volumetric flux; the rate of volume flow across a unit area [m^3 m^{-2} s^{-1}]; for example, Darcy's law of groundwater flow.

To develop your skills in modeling systems of this sort, we provide three examples. The first is representative of chemical systems, the second of sediment fluxes across the landscape, and the third of an abstract property, in this case momentum. For additional problems in diffusion, see John Crank's *The Mathematics of Diffusion* (Crank, 1980).

Translations

Example I: Dissolved Species in a Homogeneous Aquifer

Physical Picture

For our first example, consider a chemical species dissolved in the stagnant pore water of a homogeneous, isotropic aquifer. Suppose initially that it is distributed unequally throughout the aquifer, and further, assume that it is reacting with the minerals in the aquifer to form a solid phase. We want to know how the concentration varies throughout the aquifer with time.

A sketch of the aquifer (fig. 4.1) helps clarify the problem and immediately begs a number of important questions like how thick it is and whether we should consider all three dimensions. Because the aquifer is homogeneous

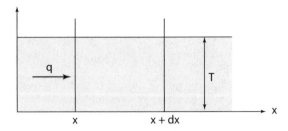

Figure 4.1. Sketch of aquifer (the shaded region in the figure). T is aquifer thickness, q is the flux of ions, and x equals distance along the aquifer.

and isotropic, we can simplify the geometry to one dimension and with no loss of generality assume it is horizontal and of thickness T [m] and unit width [m]. Next we define a control cell in which we account for the additions and subtractions of the dissolved ions. Let the left side of the cell be at x and the right side at $x + dx$, an infinitesimal distance away. Now list the important variables of the problem: Concentration of the chemical species, $C(x,t)$ [mol m^{-3}]; distance along the aquifer, x [m]; and time, t [s]. One dependent variable, C, requires one equation. Because C is a function of the independent variables x and t, the equation will be a PDE.

Physical Laws

Next we write down the laws that bear on the problem. We expect that the mass of ions of C will be conserved, and we expect that the ions will diffuse by Brownian motion according to Fick's law, named after Adolf Fick (1829–1901), a German physiologist who proposed that the following relation governs the diffusion of gases and solutes through fluids:

$$q = -D\frac{\partial C}{\partial x}, \tag{4.1}$$

where the flux of ions q has units of mol m^{-2} s^{-1}, and D is the diffusivity [m^2 s^{-1}]. (Interestingly, Fick's nephew of the same name invented the contact lens.) The rate at which the ions react with the aquifer minerals depends upon the particular reaction of course, and so for generality we will say that it is the function $S(C)$ with units mol m^{-3} s^{-1}.

Restrictive Assumptions

We are assuming that the aquifer is homogeneous and isotropic in its physical properties and its walls impermeable to fluxes of the solute. Also we assume that groundwater flow can be ignored.

Perform the Balance

Now write down the mass balance first in words:

$$TROCM = MRI - MRO + SOURCE/SINK.$$

The time rate of change of moles in the cell equals the number of moles in minus the number out. In this case we are balancing moles, not mass, but moles are equivalent to mass as long as the atomic or molecular mass of the species doesn't change. The number of moles in the control volume is the product of the concentration times the volume of the control volume. The mole rate in, representing ions entering the cell from the left, is qA, where A is the area of the aquifer face. To obtain the mole rate out, use a Taylor series expansion on the value at x. Reaction with the aquifer rock also removes the species from the cell. Thus the balance becomes in symbols:

$$\frac{\partial CAdx}{\partial t} = qA - \left(qA + \frac{\partial qA}{\partial x}dx\right) - SAdx. \qquad (4.2)$$

Check Units

After cancelling like-terms, we see that the units are consistent, which is good, but the equation contains two unknowns, C and q. These can be reduced to the one unknown C by substituting the definition of q from equation 4.1, yielding:

$$\frac{\partial C}{\partial t} - D\frac{\partial^2 C}{\partial x^2} + S = 0. \qquad (4.3)$$

Equation 4.3 is the classic 1-D diffusion equation with a sink term. *In general, any property that is conserved in one dimension and flows down a gradient according to a first-order rate law will be described by an equation similar in form to equation 4.3.* It is a second-order, parabolic PDE.

Define Interval, Specify Initial and Boundary Conditions

As defined in chapter 1, *a well-posed problem* contains as many equations as unknowns, the intervals of space and time over which we seek a solution, and initial and boundary conditions. In the problem at hand, sensible intervals are $0 < t < \infty$ and $0 < x < L$, where L is the length of the aquifer. If the aquifer is considered infinitely long, then L should be large enough such that the boundary conditions imposed at 0 and L do not appreciably influence the solution over the domain of interest. The initial and boundary conditions are necessary because as noted in chapter 1, a solution of the one-dimensional, second-order diffusion equation where the dependent variable C is a function of two independent variables (x,t) requires three functions of integration—one for time (the initial condition) and two for space (because the equation is second order in space). Let the initial condition be some function $C(x,0) = C_0(x)$. As discussed in chapter 1, the boundary conditions can be either of Dirichlet, Neumann, or mixed type. Here for example, we could use the Dirichlet boundary condition $C(0,t) = P$, where P is some temporally constant concentration, and $C(\infty,t) = 0$. It is important that your initial and boundary conditions are consistent at $x = 0$ and L.

Nondimensionalization

As noted in chapter 1, there are many advantages if we redefine the variables in equation 4.3 to make the equation dimensionless. The basic scales in this problem are length, time, and mass. We want to choose the nondimensionalizers with consideration for numerical accuracy, keeping the values around 1. Following the steps in table 1.2 of chapter 1, we choose a length scale L representing the aquifer length of interest, such that distance in the x direction is expressed in fractions of L, yielding a nondimensional length, $x^* = x/L$. To nondimensionalize time, note that the diffusivity has units of m^2 s^{-1}, so the nondimensional time could be $t^* = tD/L^2$. This allows us to measure time in terms of distances the species would diffuse in unit time. Also, one should try to incorporate the BCs into the problem. If a Dirichlet BC is defined as $C(0,t) = P$, then in the above problem it is advantageous to let $C^* = C(x,t)/P$.

Likewise, the sink term could be nondimensionalized, but to simplify the problem we drop it. Substituting the above definitions into the governing PDE (equation 4.3) and canceling terms yields:

$$\frac{\partial C^*}{\partial t^*} - \frac{\partial^2 C^*}{\partial x^{*2}} = 0. \tag{4.4}$$

Note that the equation no longer has a coefficient; one solution will satisfy all possible values of the imposed BC.

Analytic Solutions

The principle motivation for numerical modeling is the difficulty or impossibility of obtaining analytic solutions. In some cases, analytic solutions can be found, either for the original problem or for a simplification of the problem. In the latter case, the analytic solution provides an opportunity to evaluate the numerical model before applying it to the more complicated case.

There are well-known solutions to both the dimensional and nondimensional 1-D diffusion problems. For the nondimensionalized diffusion problem described above, with the initial condition $C^*(0,x) = 0$ and boundary conditions $C^*(x,0) = 1$ and $C^*(x,\infty) = 0$, the exact solution is

$$C^*(x^*,t^*) = erfc\left(\frac{x^*}{2\sqrt{t^*}}\right), \tag{4.5}$$

where *erfc* is the complementary error function. This solution has some interesting properties. The transient solution is concave up (fig. 4.2), with concentration decaying exponentially with distance. At very short times, the transient solution is unphysical, because at an infinitesimally small time greater than zero there is material everywhere. In other words, the concentration "front" travels outward at an infinite rate. At long times, the solution becomes

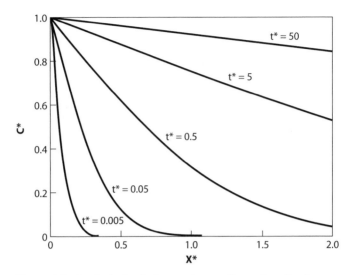

Figure 4.2. Analytical solution to the nondimensionalized 1-D diffusion equation, with $C^*(x,0) = 1$, $C^*(x,\infty) = 0$, and $C^*(0,x) = 0$.

linear, sloping from the left to the right boundary condition (which in this case is at $x = \infty$).

Example II: Evolution of a Sandy Coastline

The morphologic evolution of coastlines is a topic of considerable interest, particularly to those who own homes there. It depends upon a number of complex factors such as wave energy and refraction, tidal and coastal currents, and the balance of sediment supply and rate of creation of accommodation space. Nevertheless, some knowledge can be gained by simplifying the problem to one factor, the alongshore transfer of sediment by a wave-driven littoral current.

Physical Picture

Consider a coastline such as depicted in figure 4.3 subject to a wave field approaching at a constant angle. As these waves approach the coast obliquely and shoal in

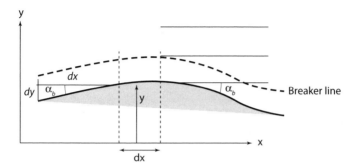

Figure 4.3. Definition sketch for example II, evolution of a sandy coastline.

the breaker zone they create a longshore current, and this current causes a longshore sediment flux. This flux redistributes sediment along the coast leading to coastal evolution. We want to know the evolution of the shoreline y as a function of location x and time t.

Physical Laws

To first order, the flux of sediment Q_s transported alongshore is linearly proportional to the longshore energy flux of the waves entering the surf zone P_{ls} (Anonymous, 1984),

$$Q_s\left(\frac{m^3}{yr}\right) = K\left(\frac{m^3 s}{N\ yr}\right)P_{ls}\left(\frac{N}{s}\right), \tag{4.6}$$

where K is a constant approximately equal to 1,290, but note that it varies as a function of sediment grain size. The longshore energy flux of waves (Joules per meter length of wave crest per unit time, or N s^{-1}) is given by

$$P_{ls} = 1.752\rho g\frac{H_{sb}^3}{T}\sin\alpha_b, \tag{4.7}$$

where ρ is the fluid density, g is gravitational acceleration, H_{sb} is the height of breaking waves in the surf zone, T is wave period, and α_b is the angle between the wave crests and the shoreline. Under the assumption that alongshore energy flux only depends upon the angle α_b, (see next section), equation 4.6 can be reduced to

$$q_s = -D\frac{\partial y}{\partial x}, \qquad (4.8)$$

where q_s is the alongshore sediment flux in m^2 s^{-1}, and D is the sediment diffusivity in m^2 s^{-1} accounting for all of the constants in equation 4.6 and equation 4.7. Thus the alongshore sediment flux takes the form of a first-order rate law.

Restrictive Assumptions

Assume α_b depends solely upon the angle of the shoreline and not on wave refraction. Also assume that H and T do not vary with time or space in this problem so that sediment flux is simply proportional to $\sin \alpha_b$. For low angles, $\sin \alpha_b$ is equal to $\tan \alpha_b$, and as indicated in figure 4.3, $\tan \alpha_b = dy/dx$. Also assume that the sediment diffusivity does not depend upon x.

Perform the Balance

To obtain an expression relating the location of the shoreline y as a function of x and t, use conservation of mass in the control volume of figure 4.3. The time rate of change of mass in the control volume equals the mass rate in minus the mass rate out. Translated into symbols, this becomes

$$\frac{\partial \rho_b y d\, dx}{\partial t} = \rho_b q_s d - \left(\rho_b q_s d + \frac{\partial \rho_b q_s d}{\partial x} dx\right), \qquad (4.9)$$

where ρ_b = the bulk density [kg m^{-3}], and T is the thickness of the active beach sediments [m]. Upon canceling terms and substituting equation 4.8 into equation 4.9, one obtains

$$\frac{\partial y}{\partial t} - D\frac{\partial^2 y}{\partial x^2} = 0. \qquad (4.10)$$

Check Units

The units are m s^{-1} for both terms.

Define Interval, Specify Initial and Boundary Conditions

To make the problem well posed, we need to define the intervals of x and t over which we seek a solution and specify initial and boundary conditions. Sensible intervals are $0 < t < \infty$ and $0 < x < L$, where L is the length of the coastal segment of interest. L should be large enough such that the boundary condition imposed there does not influence the solution over the domain of interest. Let the initial condition be some function $y(x,0) = Y(x)$. If we define the domain of interest between two points on the shoreline where $dy/dx = 0$, then by equation 4.8 the fluxes there will be zero and these become our BCs.

In summary, the evolution of a coastline can be described by a 1-D diffusion equation, although admittedly, we have undertaken some restrictive assumptions to do so. To determine whether the solutions bear any semblance to reality, try to solve modeling exercise 1 at the end of this chapter.

Example III: Diffusion of Momentum

The 1-D diffusion equation is not restricted to cases where mass in conserved; it also arises when momentum is conserved if the momentum flows according to a linear first-order rate law. Earlier, in chapter 2 we explored solutions to a 1-D diffusion equation (equation 2.26) describing the evolution of velocity profiles in a viscous fluid between two plates, one fixed and one moving at constant velocity. Here we derive that equation.

Physical Picture

Consider the situation in figure 4.4 where a reservoir of viscous fluid sits between an upper movable plate at $y = 0$ and a lower fixed plate at $y = L$. Suddenly at $t > 0$ the upper plate moves to the right at constant velocity, V_0. We want to determine the evolution of $V_x(y,t)$, where the subscript x denotes a velocity perpendicular to the y axis.

Physical Laws

The moving plate will drag along the fluid molecules immediately adjacent to it, giving the fluid an x-directed

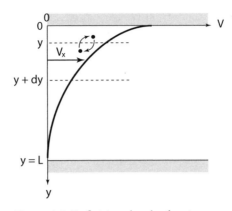

Figure 4.4. Definition sketch of a viscous fluid bounded by a ceiling that starts moving to the right at velocity V_0 at $t > 0$.

momentum mV_x, where m is the mass of a fluid parcel. Through Brownian motion, molecules in the parcel will exchange with molecules deeper in the fluid in the y direction where they will impart x-directed momentum to the surrounding molecules. Thus, x-directed momentum flows through the fluid in the y direction as long as V_x declines in the y direction. The greater the difference in velocity between any two levels in the fluid, the greater the rate of momentum flux. Thus, it is reasonable to assume that if Brownian motion is the only source of fluid interchange in the y direction, then the rate that momentum flows in the y direction, that is, its flux, is proportional to the gradient in V_x in the y direction, or

$$q_{yx} = -\mu \frac{\partial V_x}{\partial x}, \tag{4.11}$$

where q_{yx} is the flux of x-directed momentum in the y direction per unit area [kg m^{-1} s^{-2}] and μ is the proportionality constant [kg m^{-1} s^{-1}]. It is no coincidence that q_{yx} has units of force per unit area; momentum increases or decreases (i.e., the body undergoes an acceleration) as forces are applied. This interchange of molecules with different momenta and the momentum flux that accompanies it gives rise to a shearing force between the layers of fluid. In fact, equation 4.11 is just Newton's law of viscosity written to emphasize that it arises from a flux in the y direction of x-directed momentum, and q_{yx} is usually written τ_{yx}.

The second law relevant to the problem is conservation of momentum.

Restrictive Assumptions

Assume there is no fluid flow in the y direction and no body forces acting on the fluid.

Perform the Balance

To define how the velocity of the fluid varies with time and distance away from the moving plate, create a control volume of dimensions dy by unity by unity as in figure 4.4 and write down the conservation of momentum equation for that cell:

$$\text{TROCMOM}_x = \text{MOMRI}_x - \text{MOMRO}_x + \Sigma \text{Forces}_x. \tag{4.12}$$

Because we have assumed there are no body forces acting upon the cell mass, the last term is zero. Translated into symbols, equation 4.12 becomes

$$\frac{\partial \rho V_{yx} 1 \cdot 1 \, dy}{\partial t} = q_{yx} 1 \cdot 1 - \left(q_{yx} 1 \cdot 1 + \frac{\partial q_{yx} 1 \cdot 1}{\partial y} dy \right), \tag{4.13}$$

where ρ is the fluid density [kg m^{-3}]. The LHS is the time rate of change of momentum in the cell, and the RHS is the net momentum added to the cell in unit time. Upon substituting equation 4.11 into equation 4.13 and clearing terms, we arrive at

$$\frac{\partial V_x}{\partial t} - \frac{\mu}{\rho} \frac{\partial^2 V_x}{\partial y^2} = 0. \tag{4.14}$$

Notice the form is a 1-D diffusion equation.

Check Units

The coefficient of the second term is called the kinematic viscosity, ν, with units typical of a diffusivity [m^2

s^{-1}]. The units of both terms should be units of force per unit mass or acceleration, and therefore the units check.

Define Interval, Specify Initial and Boundary Conditions

To complete the problem definition we need to define the intervals of x and t over which we seek a solution and specify initial and boundary conditions. Sensible intervals are $0 < t < \infty$ and $0 < y < L$, where L is the distance between the plates. The initial condition is $V_x(y,0) = 0$, and the boundary conditions are $V_x(0,t) = V_0$ and $V_x(\infty,t) = 0$. Again, to reduce the problem to one solution covering the whole range of interest, we can nondimensionalize using $y^* = y/L$, $t^* = tv/L^2$, and $V^* = V/V_0$ to yield:

$$\frac{\partial V^*}{\partial t^*} - \frac{\partial^2 V^*}{\partial y^{*2}} = 0. \tag{4.15}$$

Finite Difference Solutions to 1-D Diffusion Problems

In chapter 2 we solved equation 4.14 to illustrate the FTCS solution scheme. Here we revisit that problem to demonstrate a solution by the Crank–Nicolson scheme. Recall that the Crank–Nicolson scheme is unconditionally stable but can suffer from large inaccuracies when the diffusion number ($s = v\,dt/dx^2$) gets large. Figure 4.5 shows the approach to steady state for the exact (analytic) solution (not visible behind the numerical simulation for $s = 25$) and for three other simulations with increasing s. With the same initial and boundary conditions as in chapter 2, the numerical solution is imperceptibly different from the analytic solution for $s < 100$ or so. As s increases beyond 500, visible oscillations appear. However, the solution oscillates about the exact value without growth in amplitude; in other words, the solution is stable, if inaccurate.

Summary

This chapter has illustrated that a large class of physical phenomena can be compactly described by the conservative

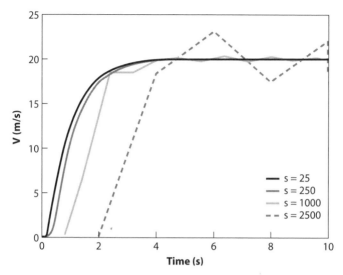

Figure 4.5. Solution to the 1-D momentum diffusion problem with boundary conditions as in chapter 2. Shown is the approach to steady state for velocity in the middle of the flow (20 mm from the plates) for four different simulations using the Crank–Nicolson scheme and different diffusion numbers (s). The solid black line is close to the analytic solution.

flow of mass or momentum down a gradient. If the physical setting is sufficiently simple so that the flow exists solely in one dimension, the resulting equation always takes the form of equation 4.15 when properly nondimensionalized. This economy of description and commonality among disparate phenomena provides great predictive power because once the properties of the solution for one phenomenon are known, they can be applied to all other phenomena.

Modeling Exercises

1. **Coastline Development**
   Given the 1-D diffusion equation derived under example II above, predict the evolution of an initial

coastline described by $y = 500 + A \sin(2\pi x/L)$, where $A = 200$ m and $L = 1,000$ m, assuming for $t > 0$ that a wave field of height $H_{sb} = 2$ m and $T = 12$ s begins striking the shore at an angle $\alpha_b = 0.2$ rad. Boundary conditions are $y(0,t) = y(2\pi,t) = 500$ m. Use any technique, analytic or numerical, that you wish.

2. **Concentration of Salt around a Dissolving Sphere**
Consider a sphere of halite of radius a, immersed in an infinite still ocean that, at $t = 0$, is freshwater. Describe the concentration of dissolved salt as a function of radial distance r, away from the sphere, and time t (1-D problem).

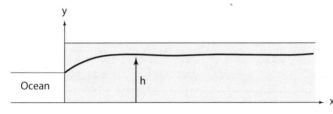

Figure 4.6. Definition sketch for modeling exercise 3. A permeable seawall at left retains a homogenous sand.

3. **Groundwater Elevation behind a Permeable Seawall**
Consider the cross section through a permeable seawall (fig. 4.6). Describe the variation in water level as a function of time and distance from the wall. Assume the sand is homogeneous with spatially and temporally constant permeability, and the water density is constant. Let the water level in the ocean vary according to a known function, $h(0,t) = H(t)$ (1-D problem).

Multidimensional Diffusion Problems

As noted in the past chapter, there is a large class of geoscience problems in which a quantity flows down a gradient according to a first-order rate law. Because that quantity is conserved, and assuming no other transport processes operate, the resulting mathematical descriptions all take the form of the diffusion equation. Gradients, of course, exist in all dimensions, and geoscientists are often faced with problems that demand a two-dimensional (2-D) or three-dimensional (3-D) approach. Here we extend the treatment to two dimensions, with examples that include the equations describing evolution of the landscape, flow in a pumped aquifer, and heat flow around a radioactive waste repository. In the latter two examples, we focus on the steady states, for which the resulting equations are classified as elliptical boundary value problems and, depending upon the problem, take one of two well-known forms—LaPlace's and Poisson's equations. For further exploration of multidimensional diffusion problems, one can turn again to Crank's *The Mathematics of Diffusion* (Crank, 1980), to Boudreau's *Diagenetic Models and Their Interpretation* (Boudreau, 1996) for geochemists, and for hydrogeologists, Hornberger and Wiberg's *Numerical Methods in the Hydrological Sciences* (Hornberger and Wiberg, 2005) is an excellent source.

Translations

Example I: Landscape Evolution as a 2-D Diffusion Problem

Consider the landscape in figure 5.1. It represents an integrated response to changing boundary conditions such as base level, climate, and land use change. Efforts to tease apart the history of Earth change recorded in landscapes has given rise to a score or more of landscape evolution models of varying complexity [see Willgoose (2004) for a thorough review]. Here we derive the simplest 2-D version possible.

Physical Picture

We seek land surface elevations above a datum, $h(x,y,t)$, starting from a known initial condition, $H(x,y,0)$, over $0 < t \leq T$. The domain of interest is the horizontal plane (x,y) where both x and y vary between 0 and L, and T and L are the duration and spatial extent of interest, respectively.

Physical Laws

From the nature of the problem, it is clear that we need one differential equation with dependent variable h as a function of three independent variables, x, y, t. To obtain the PDE, we apply the principle of conservation of mass and assume that the process of landscape development is diffusive, that is to say, mass (rock, regolith, soil) moves across the landscape at a rate proportional to topographic slope,

$$q_s = -D(s)\frac{\partial h}{\partial s}, \tag{5.1}$$

where q_s is the volumetric flux per unit width in units of $m^3 \, s^{-1} \, m^{-1}$, $D(s)$ is the diffusivity $[m^2 \, s^{-1}]$, h is the elevation at a point on the land's surface [m], and s is the horizontal axis in question (either x or y in a Cartesian coordinate system) [m].

How good is this assumption that landscapes diffuse? Studies in coastal California and the Wind River Range

Figure 5.1. Topography along the West Branch of the Susquehanna River as a diffusional landscape.

of Wyoming cited in Dietrich et al. (2003) show surprising confirmation that sediment flux down hillslopes is linear with slope, although the diffusivities vary by a factor of two, probably due to differing dominant mechanisms at the two sites. It is also true that the sediment flux of rivers is diffusive, at least to first order, as the following analysis shows.

Sediment flux in rivers is known to be a function of the bed shear stress, τ_o, and many sediment transport equations take a form similar to the classic Meyer–Peter Mueller bedload transport function (e.g., Gomez, 1991), given in dimensionless form as

$$q_s^* = 8\left(\tau_o^* - \tau_c^*\right)^{3/2}$$

where

$$q_s^* = \frac{q_s}{\sqrt{(R-1)gd^3}}$$

$$\tau^* = \frac{\tau_o}{\rho(R-1)gd}, \qquad (5.2)$$

and q_s is the volumetric sediment flux per unit width, R is the sediment density relative to water, τ_o is the bed shear stress, and τ_c is the fluid shear stress needed to initiate motion of the bed material and called the critical shear stress. It is also true that a large class of rivers maintain a constant difference between the fluid and critical shear stresses by adjusting their width [Parker (1978) as cited in Paola et al. (1992)]:

$$\tau_o - \tau_c = \tau_o\left(\frac{\varepsilon}{1+\varepsilon}\right), \qquad (5.3)$$

where ε is an empirical constant equal to about 0.4 for coarse-grained rivers. Rivers with cohesive banks typically possess bed shear stresses many times the critical shear stress of their sediment load in which case ε approaches ∞. Substituting equation 5.3 into equation 5.2 eliminates the critical shear stress from the equation to yield

$$q_s \propto \tau_o \tau_o^{1/2}. \qquad (5.4)$$

For steady, uniform flow in a hydraulically wide channel,

$$\tau_o = \rho g H S, \qquad (5.5)$$

where ρ is the fluid density, g is gravitational acceleration, H is flow depth, and S is bed slope. Substituting equation 5.5 into equation 5.4 for the first τ_o yields:

$$q_s \propto H \tau_o^{1/2} S. \qquad (5.6)$$

From the table of useful laws in chapter 1, we can make use of the quadratic shear stress relation for viscous fluids, $\tau_o = \rho C_f V^2$, where C_f is a dimensionless friction factor, and V is the average flow velocity in the vertical. Substituting this expression in equation 5.6 and noting that VH is, by definition, unit water discharge, q, we deduce that

$$q_s = -D(s)\left(\frac{\partial h}{\partial s}\right), \tag{5.7}$$

where the diffusivity $D(s)$ is given by

$$\frac{8C_f^{1/2}}{(R-1)}\left(\frac{\varepsilon}{1+\varepsilon}\right)^{3/2} q. \tag{5.8}$$

Although the friction factor and relative density are relatively constant, q scales roughly with the square root of drainage basin area. Thus, $D(s)$ can be made a known function of s. In the case of hillslopes, the diffusivity is a function of rock, type, vegetation, and climate. In rivers, it is surprising that the flux is not an explicit function of grain size. This arises because of the assumption that their widths adjust to keep the bed shear stress a fixed amount higher than the critical shear stress of their bed material.

Restrictive Assumptions

The most basic assumption in this model is that Earth materials flow across the landscape in linear proportion to the gradient in elevation. To take the simplest case, we assume that D is not a function of s (i.e., x or y). We also are treating only the case of alluvial rivers in which their slopes are fixed by the amount of load they must transport. Thus, the model applies to landscapes developed on unconsolidated materials as, for example, on coastal plains. Also, we assume that the bed material and the material in transport possess the same bulk density.

Perform the Balance

Following the rules of model-building presented in chapter 1, we define a computational cell as in figure 5.2 of volume $hdxdy$. This represents the volume of material above an arbitrary datum. Let the bulk density of the material in the cell be σ (kg m^{-3}). Now write the conservation of mass law in words:

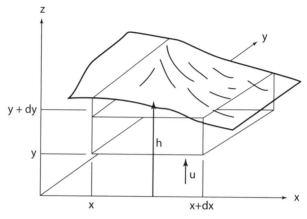

Figure 5.2. Definition sketch for a landscape evolution model. Symbols are defined in text.

Time rate of change of mass in the cell
= Mass rate in − Mass rate out. (5.9)

What are the avenues by which mass can enter this cell? At the x face, mass will flow down the slope in the x direction, entering at a rate of $\sigma q_x dy$, where q_x is the horizontal volumetric flux of sediment through the face in units of m³ per m width per unit time. At the $x + dx$ face we can use Taylor's theorem to obtain the mass flux out. Similar logic applies for the y direction. But can mass enter or exit the cell from the bottom? Yes, if there is tectonic subsidence or uplift. Denoting the velocity of vertical motion as u (positive upwards), then the mass flux through this face equals $\sigma u dx dy$. Substituting these terms into equation 5.9:

$$\frac{\partial \sigma h dx dy}{\partial t} = \sigma q_x dy + \sigma q_y dx - \left(\sigma q_x dy + \frac{\partial q_x dy}{\partial x} dx \right)$$
$$- \left(\sigma q_y dx + \frac{\partial q_y dx}{\partial y} dy \right) \quad (5.10)$$
$$+ \sigma u dx dy.$$

Upon canceling terms and substituting in equation 5.1 for q:

$$\frac{\partial h}{\partial t} - D\frac{\partial^2 h}{\partial x^2} - D\frac{\partial^2 h}{\partial y^2} - u = 0. \qquad (5.11)$$

Notice that we have taken the diffusivity out of the spatial differential, which we can only do if it does not vary with x or y. The meaning of equation 5.11 is clear if we move all terms but the first to the RHS and let the diffusivity be zero. Then the velocity of the land surface at a point is simply equal to the uplift or subsidence rate. If $u = 0$, then the landscape changes in proportion to its curvature (denoted by the second derivative, which can be thought of as a spatial change in slope).

Check Units

With D in units of $m^2\ s^{-1}$, all units are in $m\ s^{-1}$.

To simplify solutions later on, it is valuable to nondimensionalize equation 5.11 through the following transformations: $t^* = tD/L$; $x^* = x/L$; $y^* = y/L$; $h^* = h/L$. Substituting these definitions into equation 5.11 and simplifying yields

$$\frac{\partial h^*}{\partial t^*} - \frac{\partial^2 h^*}{\partial x^{*2}} - \frac{\partial^2 h^*}{\partial y^{*2}} - \frac{uL}{D} = 0. \qquad (5.12)$$

Notice that nondimensionalization has moved all the coefficients to one term that is now in the form of a Peclet number, representing the ratio of advective and diffusive mass transfer. The Peclet number will become quite important later in the book when we consider the full transport equation.

At steady state, the first term of equation 5.12 equals zero, and the resulting equation takes the form of Poisson's equation, well known to physicists and engineers. If the Peclet number equals zero, the equation takes the form of LaPlace's equation. Both then fall in the class of elliptic rather than parabolic equations, and we should expect the behavior of their solutions to reflect that fact.

Define Interval, Specify Initial and Boundary Conditions

The domain over which equation 5.12 will be solved is $0 < x^* < 1$; $0 < y^* < 1$; $0 < t^* < \infty$. A typical case is where the initial elevation is everywhere zero, the boundary nodes are always zero elevation, and at $t > 0$ the uplift rate u is specified.

Example II: Pollutant Transport in a Confined Aquifer

For our second example, consider the following typical applied hydrogeology problem. A company has inadvertently introduced a conservative pollutant into a widespread horizontal aquifer and has hired you to strategically drill a remediation well and pump it, thereby capturing the pollutant before it spreads any further. Assume that the aquifer is confined above and below by perfect aquitards. As a result, what would inherently be a 3-D problem can realistically be transformed into one that can be considered 2-D.

Physical Picture

A map of the area shows the current extent of the pollutant (projected to the surface) and some fluid heads, or heights above a datum to which water rises in a well (fig. 5.3). Of course, you want to minimize the amount of water that you treat at the remediation well head. Therefore the question becomes: Where will you place the well and what is the minimum withdrawal rate needed to accomplish your objective?

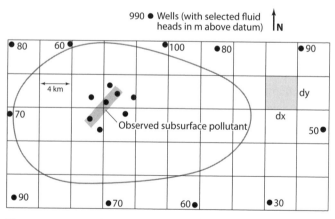

Figure 5.3. Map of study area for pollutant transport example.

Physical Laws

Groundwater flows according to a first-order rate law called Darcy's law, named after Henry Darcy (1803–1858), a French engineer whose day job was in the public works department of the city of Dijon. During his free time he quantified the loss of fluid potential due to friction (the Darcy–Weisbach equation) and performed experiments that led to what we now refer to as Darcy's law, which states that the volumetric water flux per unit area, q_s, is proportional to the gradient in fluid potential, ϕ:

$$q_s = \frac{k(s)}{\mu} \frac{\partial \phi}{\partial s}, \tag{5.13}$$

where $k(s)$ is the permeability [m^2], μ is the fluid viscosity [Pa s], ϕ is the fluid potential (Pa), and s is the horizontal distance in question (either x or y in a Cartesian coordinate system [m]). Fluid potential is defined as:

$$\phi = P + \rho g z + \frac{\rho u^2}{2}. \tag{5.14}$$

P is the fluid pressure head (equal to $\rho g h$), where ρ is the fluid density [kg m^{-3}], g is the gravitational acceleration [m s^{-2}], h is the head [m] or height to which water rises in a well, z is the elevation relative to a datum [m], and u is the fluid velocity [m s^{-1}].

Restrictive Assumptions

For the special case where fluid velocities are low, and there are negligible changes in elevation across the aquifer, $\phi = P$. Let us also assume that the porosity (α) and aquifer thickness (T), ρ, k, and μ are not functions of x, y, or t. Finally, assume that the aquifer is confined above and below by perfect aquitards.

Perform the Balance

In this problem, the volume of water in the computational cell (fig. 5.3) is $\alpha T dx dy$. The conservation of mass law written first in words is

Time rate of change of mass in the cell
    = Mass rate in − Mass rate out − Pumping   (5.15)

and then in symbols is

$$
\frac{\partial \rho \alpha T dx dy}{\partial t} = \rho \alpha q_x T dy + \rho \alpha q_y T dx
$$
$$
- \left(\rho \alpha q_x T dy + \frac{\partial \rho \alpha q_x T dy}{\partial x} dx \right)
$$
$$
- \left(\rho \alpha q_y T dx + \frac{\partial \rho \alpha q_y T dx}{\partial y} dy \right) \tag{5.16}
$$
$$
- Q \rho T dx dy.
$$

Here, Q is the withdrawal rate. The units on Q are somewhat strange, being cubic meters per second per unit volume of aquifer [s^{-1}]. It is written this way so that you may use the volume of a finite difference cell as the unit volume of reservoir. If you wanted to withdraw at 1 cubic meter per second and your grid spacing is 1,000 m on a side, then Q in equation 5.16 would be $1/(1{,}000 \times 1{,}000 \times 10)$, or 10^{-7} s^{-1}.

Dividing through by ($\alpha T dx dy$) yields:

$$
\frac{\partial \rho}{\partial t} = -\frac{\partial \rho q_x}{\partial x} - \frac{\partial \rho q_y}{\partial y} - \frac{Q\rho}{\alpha}. \tag{5.17}
$$

Now substitute in Darcy's law (equation 5.13). Because we previously assumed that ρ, k, and μ are not functions of x, y, or t, the RHS becomes zero, and we can remove them from the remaining differentials. Upon dividing through equation 5.17 by common terms, we obtain

$$
\frac{\partial^2 h}{\partial x^2} + \frac{\partial^2 h}{\partial y^2} = \frac{\mu Q}{\alpha k \rho g}, \tag{5.18}
$$

which takes the form of a Poisson equation.

Check Units
The units are m^{-1} in all three terms.

Define Interval, Specify Initial and Boundary Conditions
Let the interval be $0 < x < L$ and $0 < y < M$, where L and M are the dimensions of the study area in figure 5.3, and $0 < t < \infty$. The boundary conditions can be specified heads at the perimeter of the study area, approximated

by a (visual, numerical) interpolation from the measured heads shown in figure 5.3.

Equation 5.18 says that at steady state, the pumping term will induce a specific curvature in the heads within the aquifer. So where should one place the remediation well in the problem posed above? See the example problem at the end of this chapter where you can determine the answer yourself.

Example III: Thermal Considerations in Radioactive Waste Disposal

Physical Picture

An important occupation of geologists these days is understanding and predicting the thermal effects of radioactive waste when it is buried in geological materials. Suppose that your consulting firm has been hired by a client to locate two waste storage sites in the vertical cross-section of figure 5.4. The sites must be as close together as feasible to minimize the expense of burying the waste, yet not so close that the interfering thermal fields will cause regional

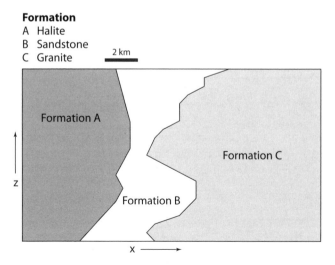

Formation
A Halite
B Sandstone
C Granite

2 km

Formation A

Formation C

Formation B

z

x

Figure 5.4. Cross section of radioactive waste repository site.

melting of the rocks. Assume that the rate of heat production from each repository is a known function of time (in units of J m^{-3} s^{-1}). Where will you place the repositories to minimize the distance between them while avoiding meltdown? Here we are conserving energy, not mass, but the derivation of the governing PDE is similar to that for mass.

Physical Laws

An obvious place to start is conservation of energy. Then assume that the heat transfer is diffusive, following Fourier's law:

$$q_s = -k(s)C\frac{\partial T}{\partial s}, \tag{5.19}$$

where q_s is the energy flux per unit width W in and out of the cross section in units of J s^{-1} m^{-2}, $k(s)$ is the thermal diffusivity [m^2 s^{-1}], C is the heat capacity of the material in the cell (J K^{-1} m^{-3}), T is the temperature [K], and s [m] is the horizontal axis in question (either x or z in a Cartesian coordinate system with x as the horizontal dimension and z in the vertical dimension).

Restrictive Assumptions

In addition to assuming Fourier's law, we also assume that advection of heat by groundwater is not important.

Perform the Balance

In this problem, we make the volume of the computational cell $Wdxdz$. Now write the conservation of energy law first in words,

> Time rate of change of energy in the cell
> = Energy rate in – Energy rate out
> + Sources – Sinks, (5.20)

then in symbols:

$$\frac{\partial}{\partial t}CTWdxdz = -\frac{\partial}{\partial x}(q_x dz Wdx)$$
$$-\frac{\partial}{\partial z}(q_z dx Wdz) + Sdxdz W. \tag{5.21}$$

Dividing through by C, W, dx, and dz, as these are not functions of time, and substituting in equation 5.19 yields

$$\frac{\partial T}{\partial t} = \frac{1}{C}\frac{\partial}{\partial x}\left(k_x C\frac{\partial T}{\partial x}\right) + \frac{1}{C}\frac{\partial}{\partial z}\left(k_z C\frac{\partial T}{\partial z}\right) + \frac{S}{C}. \tag{5.22}$$

For the special case where C and k are spatially invariant,

$$\frac{\partial T}{\partial t} = k\left(\frac{\partial^2 T}{\partial x^2} + \frac{\partial^2 T}{\partial z^2}\right) + \frac{S}{C}. \tag{5.23}$$

Check Units

All units in equation 5.22 are degrees per second as required.

Define Interval, Specify Initial and Boundary Conditions

With adequate specification of boundary and initial conditions and placement of source terms only in the repositories, equation 5.22 or equation 5.23 can be solved to establish the best location for the repositories. Solving the problem in this forward manner is not very efficient, however, because it requires guessing at two sites, running the model, and then comparing the results with other runs. Try your hand at it in the exercises at the end of the chapter.

Finite Difference Solutions to Parabolic PDEs and Elliptic Boundary Value Problems

Equation 5.12 and equation 5.23 are parabolic PDEs, and equation 5.18 is an elliptic PDE. Solutions to elliptic PDEs are governed by the boundary conditions and consequently they are called *boundary value problems* (BVPs). There exists a rich literature describing analytic solutions to elliptic BVPs, but these are generally restricted to the homogeneous case (no source or sink term), often with rather simple boundary conditions, and usually for cases with constant diffusivities. We want to avoid those assumptions and so go directly to finite difference schemes. Initially, the homogeneous case with constant diffusion coefficients is considered, but later we relax that restriction.

We start with the generic homogenous parabolic equation

$$\frac{\partial T}{\partial t} - \alpha_x \frac{\partial^2 T}{\partial x^2} - \alpha_y \frac{\partial^2 T}{\partial y^2} = 0 \qquad (5.24)$$

with Dirichlet boundary conditions

$$T(0,y,t) = a(y,t)$$
$$T(1,y,t) = b(y,t)$$
$$T(x,0,t) = c(x,t)$$
$$T(x,1,t) = d(y,t)$$

and an initial condition $T(x,y,0) = T_o(x,y)$ over the domains:

$$0 \le x \le 1$$
$$0 \le y \le 1.$$

Note that if boundary functions a, b, c, and d are not functions of time, then at large t, the system reaches steady state, that is,

$$\frac{\partial T}{\partial t} = 0,$$

and equation 5.24 reduces to LaPlace's equation, which is elliptic in nature.

An Explicit Scheme

The simplest approximation to equation 5.24 is an FTCS scheme,

$$\frac{T_{i,j}^{n+1} - T_{i,j}^{n}}{\Delta t} = \alpha_x \frac{T_{i-1,j}^{n} - 2T_{i,j}^{n} + T_{i+1,j}^{n}}{\Delta x^2} \qquad (5.25)$$
$$+ \alpha_y \frac{T_{i,j-1}^{n} - 2T_{i,j}^{n} + T_{i,j+1}^{n}}{\Delta y^2},$$

which rearranged yields

$$T_{i,j}^{n+1} = s_x T_{i-1,j}^{n} + (1 - 2s_x - 2s_y) T_{i,j}^{n} + s_x T_{i+1,j}^{n} \qquad (5.26)$$
$$+ s_y T_{i,j-1}^{n} + s_y T_{i,j+1}^{n},$$

where

$$s_x = \frac{\alpha_x \Delta t}{\Delta x^2}$$

and

$$s_y = \frac{\alpha_y \Delta t}{\Delta y^2}.$$

A Taylor series expansion about the ith, jth, nth point shows that equation 5.26 is consistent with equation 5.24 and has a truncation error of $O(\Delta t, \Delta x^2, \Delta y^2)$. A von Neumann stability analysis indicates that equation 5.26 is stable if

$$s_x + s_y \leq \frac{1}{2}. \tag{5.27}$$

These results are identical in form to the 1-D diffusion equation discussed in chapter 4. And as discussed there, in some applications this stability requirement may be overly restrictive, in which case one can move to implicit schemes that do not suffer from this restriction.

Applying this scheme to the first example in this chapter produces a landscape (fig. 5.5) that evolves as a smooth convex dome, with sediment being shed equally across all boundaries. After a sufficiently long time, a steady state is achieved where the slopes are everywhere adjusted to create the downslope sediment flux needed to just balance the sediment flux from upslope plus the mass entering the cell from uplift.

These results do not look very realistic, principally because we have made no attempt to accumulate flow as a function of an evolving upstream drainage basin area. To do so requires letting the diffusivities be a function of x and y.

Implicit Schemes

FTCS Fully Implicit

The next most obvious scheme uses approximations that are also forward in time and centered in space, but instead of approximating the spatial derivatives at the

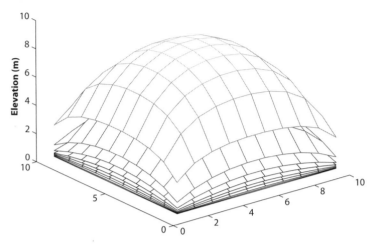

Figure 5.5. Simulated landscape under constant uplift rate and boundaries fixed at zero.

current time step, they are approximated at the new time step. This results in a system of equations, one for each node in the computation domain, providing just enough equations as there are unknowns.

The FTCS fully implicit approximation to equation 5.24 is

$$s_x T^{n+1}_{i-1,j} - (1 + 2s_x + 2s_y) T^{n+1}_{i,j} + s_x T^{n+1}_{i+1,j} + s_y T^{n+1}_{i,j-1} + s_y T^{n+1}_{i,j+1} = -T^n_{i,j}, \quad (5.28)$$

where s_x and s_y are defined as before. The computation module is given in figure 5.6. This scheme has a truncation error of $O(\Delta t^2, \Delta x^2, \Delta y^2)$, which is better than the explicit scheme and furthermore is unconditionally stable. But as noted earlier, the time step is still constrained by considerations of accuracy.

As an example application of algebraic equation 5.28, consider the problem of determining the values of $T(2,2)$ and $T(3,2)$ in figure 5.7. This example is similar to the 1-D example given in chapter 2. Writing equation 5.28 twice, once for node $(2,2)$ and once for $(3,2)$, yields

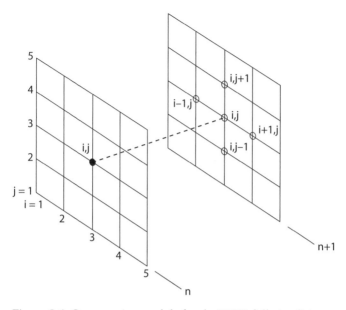

Figure 5.6. Computation module for the FTCS fully implicit scheme. [Modified from Hoffmann, K. A., and S. T. Chiang (2000). *Computational Fluid Dynamics for Engineers.* Wichita, KS, Engineering Education System.]

$$-s_x b + (1 + 2s_x + 2s_y)\, T_{2,2}^{n+1} - s_x T_{3,2}^{n+1} - s_y d - s_y m = T_{2,2}^n$$
$$-s_x T_{2,2}^{n+1} + (1 + 2s_x + 2s_y)\, T_{3,2}^{n+1} - s_x g - s_y e - s_y l = T_{3,2}^n \quad (5.29)$$

or in matrix notation:

$$
\begin{pmatrix} (1 + 2s_x + 2s_y) & -s_x \\ -s_x & (1 + 2s_x + 2s_y) \end{pmatrix} \begin{pmatrix} T_{2,2}^{n+1} \\ T_{3,2}^{n+1} \end{pmatrix}
$$
$$
= \begin{pmatrix} T_{2,2}^n + s_x b + s_y (d + m) \\ T_{3,2}^n + s_x g + s_y (e + l) \end{pmatrix}. \quad (5.30)
$$

There are just enough equations as there are unknowns and thus matrix algebra can be used to solve for the T vector of unknowns if the initial conditions $T_{i,j}^n$ are known. Although this example of two unknowns makes it appear

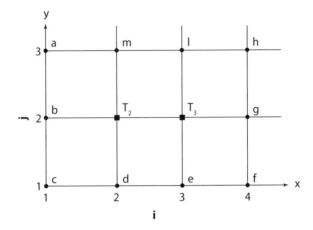

Figure 5.7. Example finite difference grid illustrating the FTCS fully implicit scheme. Values of the dependent variable are to be calculated at nodes (2,2) and (3,2) given information on the boundaries of the grid where $T(1,3) = a$, $T(1,2) = b$, and so forth.

that the terms in the A coefficient matrix are grouped along the diagonal, in fact for a larger number of unknowns the matrix is pentadiagonal, making its inversion more computationally intensive. One approach to a more efficient solution is the alternating direction implicit (ADI) method.

Alternating Direction Implicit Method

As noted in chapter 2, a tridiagonal matrix can be inverted rapidly using Thomas' algorithm. This fact is exploited in the ADI method, by subdividing the time step into 2 half time steps. In the first half time step, $T_{i,j}$ is approximated by

$$
\frac{T_{i,j}^{n+1/2} - T_{i,j}^{n}}{\frac{\Delta t}{2}} = \alpha_x \frac{T_{i-1,j}^{n+1/2} - 2T_{i,j}^{n+1/2} + T_{i+1,j}^{n+1/2}}{\Delta x^2}
$$

$$
+ \alpha_y \frac{T_{i,j-1}^{n} - 2T_{i,j}^{n} + T_{i,j+1}^{n}}{\Delta y^2}. \tag{5.31}
$$

The set of algebraic equations is implicit in the x direction, whereas old values of T are used in the approximation of the curvature in the y direction. Because only nodes $i - 1$, i, and $i + 1$ are used, the resulting coefficient matrix is tridiagonal. In the second half time step, the set of equations is implicit in the y direction,

$$
\frac{T_{i,j}^{n+1} - T_{i,j}^{n+1/2}}{\frac{\Delta t}{2}} = \alpha_x \frac{T_{i-1,j}^{n+1/2} - 2T_{i,j}^{n+1/2} + T_{i+1,j}^{n+1/2}}{\Delta x^2}
$$
$$
+ \alpha_y \frac{T_{i,j-1}^{n+1} - 2T_{i,j}^{n+1} + T_{i,j+1}^{n+1}}{\Delta y^2}. \tag{5.32}
$$

whereas the values updated to the half time step are used to approximate the curvature in the x direction. Again, the coefficient matrix of the equation set is tridiagonal. Although a penalty is paid by inverting two matrices per time step, overall computational efficiency is nevertheless increased.

Case of Variable Coefficients

If the diffusivities are variable, we must go back to equation 5.10, substitute in equation 5.7, and apply the product rule of calculus:

$$
\frac{\partial h}{\partial t} = \frac{\partial (D \frac{\partial h}{\partial x})}{\partial x} + \frac{\partial (D \frac{\partial h}{\partial y})}{\partial y} + u
$$
$$
= D \frac{\partial^2 h}{\partial x^2} + D \frac{\partial^2 h}{\partial y^2} + \frac{\partial D}{\partial x} \frac{\partial h}{\partial x} + \frac{\partial D}{\partial y} \frac{\partial h}{\partial y} + u. \tag{5.33}
$$

Notice that new terms arise reflecting the variation of D with x and y. Interestingly, these terms are multiplied by the slopes in the x and y directions, reflecting the fact that variation of D in x and y only matters if there are height gradients in the x and y directions. These new terms change the form of the PDE, because there are first- and second-order terms in the spatial derivatives. The equation now has the form of an advection–diffusion equation, discussed in detail in chapter 7, with the "velocity" equal to the spatial gradient in D.

In equation 5.33, we expanded the diffusion terms to show that the diffusivities are functions of x and y. Alternatively, we could retain the diffusivities inside the partials as in

$$\frac{\partial \left(D \frac{\partial h}{\partial x} \right)}{\partial x}, \tag{5.34}$$

but how should we treat such terms in a finite difference scheme? One common approach is to estimate their magnitude at a point in the grid particularly appropriate for the spatial gradient they are multiplying. For example, if the curvature at point i is estimated by a centered difference that subtracts a slope taken behind point i from one taken ahead, then the diffusivities should be located at the half-space steps centered on their specific gradients:

$$\frac{D_{i+1/2,j} \frac{\left(T_{i+1,j}^n - T_{i,j}^n \right)}{\Delta x} - D_{i-1/2,j} \frac{\left(T_{i,j}^n - T_{i-1,j}^n \right)}{\Delta x}}{\Delta x}. \tag{5.35}$$

One way to do this is to use averages, as, for example,

$$D_{i+1/2,j} = \frac{D_{i+1,j} + D_{i,j}}{2}. \tag{5.36}$$

Summary

We've seen that expanding our consideration of diffusive problems to a second dimension has opened up a rich variety of model derivations requiring new methods of numerical solution. We've also seen that some problems that do not initially appear to have diffusive rate laws (e.g., river transport of sediment) simplify to diffusive problems when reasonable constitutive relations are applied. We learned that if diffusivities are homogeneous in space, approaches we developed for the solution of one-dimensional diffusion problems can be applied. The fully implicit method, however, involves a pentadiagonal matrix that is computationally intensive to invert. In such cases, the ADI method is often employed.

Modeling Exercises

1. Pollutant Transport in a Confined Aquifer

Consider a special application of the "Pollutant Transport in a Confined Aquifer" problem presented earlier in this chapter. The aquifer is 10 m thick, with known rock permeability, $k = 10^{-11}$ m^2 (about 10 darcy), and the water viscosity $\mu = 10^{-3}$ Pa s. To make the problem interesting, assume that you may *not* place your well within the area demarcated by the solid elliptical line in figure 5.3. Where will you place your well to collect all of the contaminant with the smallest pumping rate?

2. Radioactive Waste Disposal

Refer back to the "Radioactive Waste Disposal" example earlier in the chapter. Recall that your task is to locate two waste storage sites in the vertical cross-section of figure 5.4. The sites must be as close together as feasible to minimize the expense of burying the waste, yet not so close that the interfering thermal fields will cause regional melting of the rocks. The repositories will each be 1,000 m^3 in volume. Your client tells you that the rate of heat production from each will be given by the equation

$$S = S_0 e^{-\lambda t}, \tag{5.37}$$

where $S_0 = 8 \times 10^{-3}$ J m^{-3} s^{-1} and $\lambda = 0.0014$ y^{-1}. Furthermore, materials will not be placed in the sites at the same time, but offset by 100 years. Assume in figure 5.4 that all geological boundaries are sharp. Assume reasonable BCs and ICs. Where will you place the repositories to minimize the distance between them while avoiding meltdown? Given the spatial heterogeneity of the lithology of the subsurface, you should find a solution to equation 5.22. To do so will require special attention be paid to cells along the boundaries of the various lithologies. Values for k and C for the various lithologies may be found in the literature.

3. Dispersion of a Pollutant

Fifty kilograms of a nonreactive pollutant is introduced into the center of a still, roughly rectangular flooded strip mine 1 km in length, 100 m wide, and 20 m deep, with precipitous sides that drop almost immediately to 20 m. The pollutant is quickly dispersed as a strip 5 m wide across the width of the strip mine and to a depth of 1 m in the center of the water body (500 m from either end). Calculate the spread of this pollutant over the next 10 days, given a horizontal diffusivity of 0.10 m^2 s^{-1} and a vertical diffusivity of 5×10^{-6} m^2 s^{-1}. Treat this as a 2-D problem, with one axis along the length of the water body and the other from the surface to the bottom.

Advection-Dominated Problems

In this chapter, we consider another common process that transports mass and momentum into and out of geological reservoirs—passive transport by the motion of a medium such as water or air. The transport of a conserved property by a fluid in motion is called *advection* or *convection*. Although the terms are often used synonymously, convection is understood in some disciplines to mean the total transport of a substance by both diffusive and advective processes, whereas in others it is the transport of a substance by combined molecular and eddy diffusion as opposed to macroscopic fluid flow. Yet other disciplines such as ocean and atmospheric sciences think of advection as a transport mechanism associated with the mean flow (largely horizontal) and convection as a largely vertical transport associated with buoyancy contrasts. Here we will use both terms to mean the passive transport of a substance by flow of the medium in which the substance is contained. Geological situations in which advection or convection is important include the transport of dissolved species in groundwater and surface waters, heat in lava flows, and suspended sediment in rivers. The fluid motion is represented by a vector, and the property or substance being transported is represented by a scalar quantity. Because we want to emphasize certain basic concepts, the focus will be on one-dimensional problems. For a more in-depth treatment of advection, the reader is referred to Fletcher (1991) or Anderson (1995).

Translations

Example I: A Dissolved Species in a River

When a pollutant is discharged into a river, the pollutant is carried passively or advected downstream by the flowing water in addition to spreading by diffusion. Advection is typically many times faster than diffusion in most rivers, and if approximate answers suffice, then diffusion can be ignored. As an example of this class of problem, consider the river reach in figure 6.1. Often, we would like to predict the time of arrival of the pollution front and the subsequent time history of its concentration, c (kg m^{-3}) at various locations downstream, assuming variations of concentration within the cross section at a location are not important to us. For an application of this technique to measure stream flow, see Herschy (2009).

Physical Picture

Let the dependent variable c be a function of the independent variables distance along stream x and time t. Under the assumptions of this 1-D problem, the control cell can be defined as a volume dx [m] long with a

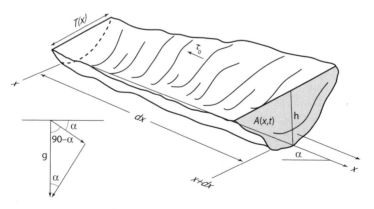

Figure 6.1. Definition sketch of a river reach carrying a pollutant.

cross-sectional area, A [m^2]. Let the speed of the river be a cross-sectional average velocity u [m s^{-1}] that varies with x and t. For the model to be generally applicable, A also will vary with x. Note that the water discharge Q [m^3 s^{-1}] equals uA by definition. In summary, we want a function for $c(x,t)$ over the intervals $0 \le x \le L$ and $0 \le t \le T$ (where L is the reach length of interest and T is the time period of interest) and subject to initial and boundary conditions yet to be specified.

Physical Laws

An obvious place to start is conservation of mass.

Restrictive Assumptions

Assume that one dimension is adequate for the intended purpose and the pollutant is passively carried by the flow with no dispersion.

Perform the Balance

In words,

$$\text{TROCM}_p = \text{MRI}_p - \text{MRO}_p, \tag{6.1.}$$

and then in symbols:

$$\frac{\partial cA dx}{\partial t} = cQ - \left(cQ + \frac{\partial cQ}{\partial x} dx \right). \tag{6.2}$$

The LHS is the time rate of change of the mass of pollutant in the control cell, which is given by its concentration in the control cell times the volume of the cell. On the RHS, the mass rate into the cell through the river cross section at x is given by the volumetric flux of water (Q) times the mass of pollutant per unit volume (c). The mass flux out at $x + dx$ is defined by Taylor series (remember equation 1.14). Clearing common terms that can be taken out of the differentials yields

$$\frac{\partial cA}{\partial t} = -\frac{\partial cQ}{\partial x}. \tag{6.3}$$

Using the product rule and grouping derivatives with like coefficients yields

$$c\left(\frac{\partial A}{\partial t} + \frac{\partial Q}{\partial x}\right) + A\frac{\partial c}{\partial t} + Q\frac{\partial c}{\partial x} = 0. \tag{6.4}$$

This can be simplified by taking advantage of the fact that water also is conserved in the reach. Performing a mass balance on water in the reach,

$$\text{TROCM}_w = \text{MRI}_w - \text{MRO}_w \tag{6.5}$$

$$\frac{\partial \rho A dx}{\partial t} = \rho Q - \left(\rho Q + \frac{\partial \rho Q}{\partial x} dx\right) \tag{6.6}$$

or

$$\frac{\partial A}{\partial t} + \frac{\partial Q}{\partial x} = 0, \tag{6.7}$$

where ρ is the density of water [kg m^{-3}]. Thus, the first term in equation 6.4 is identically zero. After dividing through by A, equation 6.4 becomes

$$\frac{\partial c}{\partial t} + u\frac{\partial c}{\partial x} = 0. \tag{6.8}$$

Check Units

The units are correctly balanced because each term has units of kg m^{-3} s^{-1}.

Define Interval, Specify Initial and Boundary Conditions

As noted earlier, first-order PDEs require one function of integration for each independent variable. These are obtained from the initial and boundary conditions we provide, such as $c = C_1(x,0)$ and $c = C_2(0,t)$, respectively. The specific functions C_1 and C_2 depend upon the specific problem of course, and some examples will be given later.

What class of PDE is this? Remember from chapter 1 (see equation 1.11) that the class depends upon the sign of $B^2 - 4AC$ where A and so forth are the coefficients of a general second-order, linear PDE. Upon first inspection it appears equation 6.8 is unclassifiable by this method. But if one takes the time derivative of both sides and assumes that u does not vary with time,

$$\frac{\partial^2 c}{\partial t^2} + u \frac{\partial}{\partial t} \frac{\partial c}{\partial x} = 0, \qquad (6.9)$$

one sees by comparison with equation 1.10 that $A = 1$, $B = u$, $C = D = E = F = 0$, and consequently $B^2 - 4AC > 0$. Therefore, equation 6.8 is a first-order, homogeneous, hyperbolic PDE. It is sometimes called the one-way wave equation.

Hyperbolic equations like equation 6.8 can be solved analytically by the method of separation of variables or the method of characteristics. For the purposes here, let us guess at a specific solution. What simple function has a time derivative that will equal u times its spatial derivative? One answer is

$$c = x - ut, \qquad (6.10)$$

because upon substitution in equation 6.8, $-u + u = 0$ as required. What initial and boundary conditions have we unconsciously assumed? Letting $t = 0$ shows that $c = x$ (i.e., the initial condition is a concentration that increases linearly from $x = 0$ to ∞ and everywhere is equal to the value of x itself). What is the BC at $x = 0$? It is $c = -ut$, which is to say that through time, c becomes increasingly negative at a rate of u. Neither of these represents a common geological condition, but both serve our purpose here, which is to show the behavior of the advection equation. What is the solution through time? It is an advancing front as shown in figure 6.2, and it is generally true for all solutions of equation 6.8 that an initial concentration profile will be advected unchanged at the speed u.

This behavior is readily apparent if one remembers that by definition

$$u = \frac{dx}{dt}, \qquad (6.11)$$

and therefore equation 6.8 is equivalent to the ordinary differential equation

$$\frac{dc}{dt} = 0 \quad \text{along the particle path} \quad \frac{dx}{dt} = u. \qquad (6.12)$$

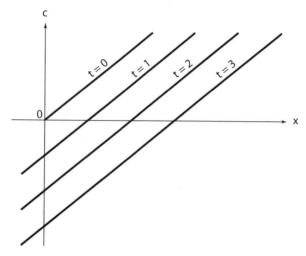

Figure 6.2. Solution of equation 6.8 for an initial and boundary condition of $c(x,0) = x$ and $c(0,t) = -ut$, respectively.

Thus, $c(x,t)$ is unchanged in time, and the initial conditions are advected with the flow without any change of form. This conclusion also points out an important point about the boundary conditions for hyperbolic problems. Because information is translated across the domain from boundaries, it is important to specify the correct number and location of boundary conditions.

In the above example, u was a constant and not a function of the concentration of pollutant, space, or time. But often in geological problems, u depends upon the dependent or independent variables, thereby making analytic solutions difficult to obtain. The following example illustrates these more typical cases.

Example II: Lahars Flowing along Simple Channels

The Indonesian term for mixtures of water and pyroclastic debris flowing down the slopes of a volcano is *lahar*. Flowing wet concrete provides a good analogue. A particularly well-studied location for lahars lies on the slopes of Mt.

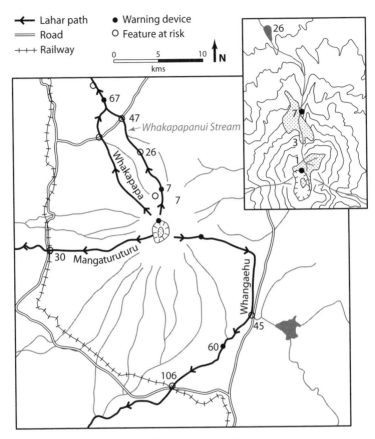

Figure 6.3. Vicinity of Mt. Ruapehu, New Zealand, showing potential path of lahars along Whakapapanui Stream. [Modified from Houghton, B., et al. (1987). Volcanic hazard assessment for Ruapehu composite volcano, Taupo volcanic zone, New Zealand. *Bulletin of Volcanology* 49(6):737–751.]

Ruapehu on the North Island of New Zealand (fig. 6.3). The first published observation occurred in 1859 (Vignaux and Weir, 1990); the largest observed lahar, in 1953, swept out a railroad bridge (site 106) right before the arrival of the Wellington–Auckland express, killing 151 people. Predicting the character of lahars that might sweep down the various valleys of Mt. Ruapehu is of obvious interest. Based on observations of 13 lahars between 1953 and 1977, the volumes range from 18,000 to 1,900,000 m³ with an average of 500,000 m³ (Vignaux and Weir, 1990). For the purposes of this exercise, assume that a lahar originating at the warning device on the crater rim (fig. 6.3) travels along Whakapapanui Stream through a rectangular valley everywhere 50 m wide and of constant bed slope S. We want to know the travel time to the road and the maximum depth and discharge expected there.

Physical Picture

Given that we are only interested in the down-valley motion of a constant-width lahar, the problem reduces to one dimension. A definition sketch is given in figure 6.4, wherein q is the volumetric flow rate or discharge per unit width of the lahar [m² s⁻¹] equal to vh, v is the average velocity in the vertical [m s⁻¹], h is the lahar thickness [m], and σ is the bulk density [kg m⁻³].

Physical Laws

The two dependent variables would seem to be q (or v) and h, and these are functions of the independent variables x and t, as well as bed slope S, and whatever other parameters come to bear on the problem. Two dependent variables require two equations. In problems like this where a mass has a velocity, it is always safe to call upon the

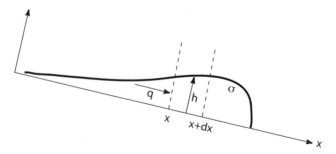

Figure 6.4. Definition diagram for the motion of a lahar. See text for details.

equations describing conservation of mass and conservation of momentum.

Restrictive Assumptions

So far we have assumed that (1) the lahar is of constant width such that a 1-D balance is adequate; (2) the lahar does not entrain material along its path; and (3) the lahar rheology is similar to that of water.

Perform the Balance

The law of conservation of lahar mass in words is

$$\text{TROCM} = \text{MRI} - \text{MRO}. \qquad (6.13)$$

The mass in the control cell is the cell volume $B\,h\,dx$ times the bulk density of the lahar, where B is unit width [m]. The volumetric rate at which lahar material enters the cell through the face at x is the volumetric flux per unit width, q, times the width. This is converted to a mass flux when multiplied by the bulk density σ. The mass rate out of the cell is obtained as usual by Taylor series, yielding

$$\frac{\partial \sigma Bh dx}{\partial t} = \sigma q B - \left(\sigma q B + \frac{\partial \sigma q B}{\partial x}dx\right) \qquad (6.14)$$

or, if the width and bulk density do not vary in space or time,

$$\frac{\partial h}{\partial t} + \frac{\partial q}{\partial x} = 0. \qquad (6.15)$$

Now we need a second equation that relates some combination of q, h, x, and t. This can be obtained by considering a force balance on a segment of the lahar. Consider figure 6.1 again, but assume that the channel cross-section is filled by a lahar. If the lahar is flowing at constant velocity, then the downslope component of gravity acting on the lahar mass between x and $x + dx$ must be balanced by a frictional retarding force acting on the lahar surface in contact with the bed and banks. The downslope gravity force is given by

$$\sigma A dx\, g \sin \alpha, \qquad (6.16)$$

where A is the cross-sectional area of the flow, and, for small bed slopes, sin α equals tan α equals the bed slope S. The retarding force is the tangential shear stress between the flow and the bed and banks times the area over which it acts, $\tau_o P dx$, where P is the wetted perimeter, which for a rectangular channel equals the flow width plus two times the flow depth. Equating the two yields

$$\tau_o = \sigma g R S, \tag{6.17}$$

where A/P is called R, the hydraulic radius. We will assume here that the flow width of the lahar is more than 20 times the flow depth, in which case the hydraulic radius is not sensibly different from the flow depth. Now recall from chapter 1 that a turbulent flow of water or air exerts a shear stress on its boundaries proportional to the square of its depth-averaged velocity, or

$$\tau_o = C_f \sigma v^2, \tag{6.18}$$

where C_f is a coefficient of drag. The desired relationship between the vertically averaged flow velocity and flow depth is obtained by equating the previous two equations, yielding

$$v = \beta \sqrt{h}, \tag{6.19}$$

where

$$\beta = \sqrt{\frac{gS}{C_f}},$$

and we have assumed $R = h$, the flow depth. Equation 6.19 is a generalization of the Chézy equation, named after Anton Chézy, a French engineer who derived it in 1768. It predicts the cross-sectional average velocity of steady, uniform flows.

Substituting equation 6.19 into the definition of q yields

$$q = \beta h^k, \tag{6.20}$$

where β depends upon local bed slope and roughness, and $k = 1.5$. Application of equation 6.20 to observed lahars indicates that k actually ranges between 1.24 to 1.47 (Vignaux and Weir, 1990), but we retain 1.5 for simplicity.

To reduce the number of dependent variables to one, substitute equation 6.20 into equation 6.15 to yield

$$\frac{\partial h}{\partial t} + \frac{\partial \beta h^k}{\partial x} = 0, \qquad (6.21)$$

or upon expanding,

$$\frac{\partial h}{\partial t} + \beta k h^{k-1} \frac{\partial h}{\partial x} = 0. \qquad (6.22)$$

We have accomplished our objective of deriving an equation that predicts the variation in lahar thickness as a function of time and space. Notice that it takes the form of an advection equation.

Check Units

As noted earlier, the units on β are $m^{1/2}$ s^{-1}. Thus, it is important to note that the units are only correctly balanced when $k = 1.5$.

Define Interval, Specify Initial and Boundary Conditions

Let the domain be $0 < x < L$ and $0 < t < t_{max}$. The initial condition of the lahar can be approximated two ways. We can define its lateral extent and thickness as an initial condition, such as $h = h_o$ for $0 \leq x \leq X$ and $h = 0$ for $X < x \leq L$, where X is the initial lateral extent. In this case, the boundary condition would be $h = 0$. Alternatively, the initial condition can be $h(x,0) = 0$, and then the lahar is introduced as a boundary condition through a known function, $h = h_o(0,t)$. The constant β depends upon the bed slope along the potential lahar path and the coefficient of drag. Inspection of a topographic map shows an average slope to be about 0.1. Typical C_f values for river channels are about 10, and thus β is about 1/3.

Notice that equation 6.22 differs from the first example (equation 6.8) in that the velocity of propagation is now a function of the dependent variable, h, increasing as the thickness of the lahar increases. This makes the equation nonlinear and leads to a rich behavior as the following thought experiment demonstrates. Consider a lahar whose surface at $t = 0$ may be described by a half cycle of a sine wave. Upon propagation, the crest of the lahar will travel faster than the trough, thereby steepening the crest front and shallowing the slope of the back limb leading to a breaking wave and a shock front. Equations of this form are treated more thoroughly in chapter 8.

Finite Difference Solution Schemes to the Linear Advection Equation

Although analytic solutions to particular forms of the advection equation can be found, it is usually simpler to obtain finite difference solutions. But even finite difference solutions are tricky because the shocks they can produce are always difficult for numerical schemes to handle. Moreover, although stable schemes are available, they have numerical issues including the appearance of diffusion-like behavior and wakes that compromise accuracy.

We begin the discussion by considering a linear hyperbolic PDE of the general form

$$\frac{\partial h}{\partial t} + a\frac{\partial h}{\partial x} = f(x,t), \tag{6.23}$$

where a is a constant advection speed (independent of h), and $f(x,t)$ is a source or sink term. In the case of a lahar, it could account for addition of sediment from the bed of the channel. The discretization of equation 6.23 is fairly straightforward using the finite difference operators defined in chapter 2 and the definitions

$$t^n = t^0 + n\Delta t \quad x_i = x_0 + i\Delta x, \tag{6.24}$$

where $n = 1, 2, 3...N$ and $i = 1, 2, 3...L/\Delta x$.

Four common schemes are the following:

FTCS

$$\frac{h_i^{n+1} - h_i^n}{\Delta t} + a\frac{h_{i+1}^n - h_{i-1}^n}{2\Delta x} = f_i^n + O(\Delta t, \Delta x^2). \tag{6.25}$$

Upwind

$$\frac{h_i^{n+1} - h_i^n}{\Delta t} + a\frac{h_i^n - h_{i-1}^n}{\Delta x} = f_i^n + O(\Delta t, \Delta x) \quad (a > 0). \tag{6.26}$$

Multistep Method (Leapfrog)

$$\frac{b_i^{n+1} - b_i^{n-1}}{2\Delta t} + a\frac{b_{i+1}^n - b_{i-1}^n}{2\Delta x} = f_i^n + O(\Delta t^2, \Delta x^2). \quad (6.27)$$

Crank–Nicolson Implicit

$$\frac{b_i^{n+1} - b_i^n}{\Delta t} + \frac{a}{2}\left[\frac{b_{i+1}^{n+1} - b_{i-1}^{n+1}}{2\Delta x} + \frac{b_{i+1}^n - b_{i-1}^n}{2\Delta x}\right]$$
$$= \frac{f_i^{n+1} + f_i^n}{2} + O(\Delta t^2, \Delta x^2). \quad (6.28)$$

Table 6.1 shows that the FTCS scheme is not stable, and it will not be discussed further here. All of the other schemes are stable, although some stability restrictions are more stringent than others. Also remember from chapter 2 that stability is a necessary but not sufficient condition for accuracy.

An important necessary (but not sufficient) stability criterion for advective problems involves the Courant–Friedrichs–Lewy parameter C, also called the *Courant number* and defined as

$$C \equiv \frac{u\Delta t}{\Delta x}.$$

It gives the fraction of a space step that a signal has traveled during a time step. If it becomes greater than 1, information about the dependent variable completely bypasses a node, causing instability.

The accuracy of the three stable schemes can be explored by comparing their solutions to an analytic solution.

Table 6.1. Four Common Finite Difference Schemes for the Advection Equation

Method	Type	Stability requirements[a]
FTCS	Explicit one-step	Unconditionally unstable
Upwind	Explicit one-step	Stable for $0 < |C| \leq 1$
Leapfrog	Explicit multistep	Stable for $|C| \leq 1$
Crank–Nicolson	Implicit one-step	Stable

[a]C = Courant number.

The comparison is facilitated by dropping the source/sink term and nondimensionalizing equation 6.23 using the following definitions:

$$h^* = \frac{h}{h_o}; \quad x^* = \frac{x}{L}; \quad t^* = \frac{t}{t_{max}},$$

where the nondimensionalizers come from our boundary conditions and domain definitions. Then the nondimensional advection speed is given by

$$\alpha = a\frac{t_{max}}{L}, \tag{6.29}$$

and equation 6.23 becomes

$$\frac{\partial h^*}{\partial t^*} + \alpha\frac{\partial h^*}{\partial x^*} = 0. \tag{6.30}$$

Let $\alpha = 1$, with initial conditions

$$\begin{aligned} h^*(x,0) &= \sin(10\pi x^*) \quad 0 \le x^* \le 0.1 \\ h^*(x,0) &= 0 \quad\quad\quad 0.1 < x^* \le 1 \end{aligned} \tag{6.31}$$

and boundary conditions: $h^*(0,t^*) = 0$ and $h^*(1,t^*) = 0$. For comparison, the analytic solution in the interior of the domain away from both boundaries is

$$h^*(x^*,t^*) = \sin(10\pi[x^* - \alpha t^*]). \tag{6.32}$$

When solutions from the upwind scheme are compared with the analytic solution (fig. 6.5), it is apparent that the scheme is exact if the Courant number $C = 1$ but produces an artificial numerical diffusion at lower values of C. The origin of this diffusion is apparent if the upwind scheme is tested for consistency. First, expand h_i^{n+1} and h_{i-1}^n using Taylor series and substitute the results in equation 6.26. Then cancel terms and realize that

$$\frac{\partial^2 h}{\partial t^2} = -a^2\frac{\partial^2 h}{\partial x^2}.$$

This results in

$$\frac{\partial h}{\partial t} + a\frac{\partial h}{\partial x} - \frac{a\Delta x}{2}(1-C)\frac{\partial^2 h}{\partial x^2} + O(\Delta x^2, \Delta t^2) = 0. \tag{6.33}$$

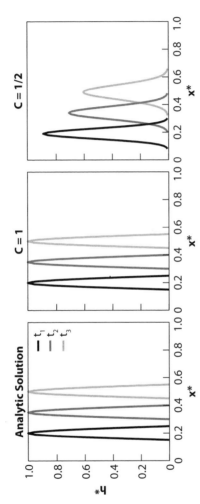

Figure 6.5. Solutions from an upwind scheme compared with an analytic solution at three times. If $C = 1$, the solution is exact; with decreasing C, the solution shows increasing numerical diffusion.

Thus the upwind scheme is not consistent with the advection equation; it contains a diffusion term. It is accurate when $C = 1$ only because the diffusion term reduces to zero.

The leapfrog scheme is stable if $C < 1$, but it requires a starter solution because a value at $n - 1$ is needed, making it less desirable. The Crank–Nicolson scheme would seem to be ideal because it is unconditionally stable and of second-order accuracy. But as figure 6.6 shows, the Crank–Nicolson scheme has problems. The solution travels more slowly than it should, lagging behind the exact solution, and a train of dispersive waves trails behind.

The case where the coefficient a above depends upon x or t requires particular care because a solution scheme may be stable in one part of x–t space and not in another.

Summary

When a substance or property is passively convected along in a flowing fluid, its evolving properties can be described by a first-order hyperbolic equation called the advection or convection equation. When the speed of the advection depends upon the magnitude of the substance or property, the equation becomes nonlinear, and shock fronts and other interesting features can arise in the solutions. The advection equation is difficult to solve accurately by finite difference. The simplest, reasonably accurate scheme is the upwind scheme with $C = 1$, but even it is not a universal solution because geophysical flows often reverse direction, and it is not always possible to know which direction is "upwind." Also, C often is a function of x, and therefore it is difficult to keep $C = 1$. In chapter 9, we will return to the nonlinear advection equation and provide more robust solution schemes.

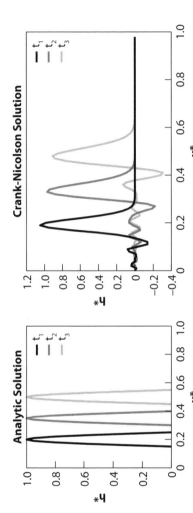

Figure 6.6. Solution to the linear advection equation using the Crank–Nicolson implicit scheme with $C = 0.5$. A wake has formed behind the primary wave.

Modeling Exercises

1. **Concentration of a Dissolved Species in a Confined Reservoir**

   A confined aquifer of cross section A [m^2] is passing a dissolved constituent in the x direction that is reacting with the aquifer mineral grains at a rate proportional to its concentration. Take the groundwater flow rate as U [m s^{-1}]. Describe the concentration of the constituent as a function of horizontal distance and time (1-D problem). Define your own initial and boundary conditions.

2. **Sedimentation of a Surface Signal**

   A 1-cm layer of contaminated sediment is deposited on the bottom of a deep lake. The contaminant concentration C is 100 mg g^{-1} dry sediment. Background sedimentation is 0.5 cm y^{-1}, and the background contaminant concentration is 0.1 mg g^{-1} dry sediment. Assume that the contaminant is immobile and that there is no bioturbation or other disturbance of the sediment. Calculate the concentration profile (with the sediment–water interface as $x = 0$ cm) for 5 years, 50 years, and 500 years after the contamination event, assuming:

   a. constant porosity, $\phi = 0.8$, and no reactivity of the contaminant;

   b. decreasing porosity according to $\phi = a \exp(-bx) + c$, where, if x is expressed in cm, $a = 0.8$, $b = 0.01$, and $c = 0.6$;

   c. decreasing porosity as in (b) and a first-order decay rate for the contaminant of $R = -0.05$ (per year) C.

3. **River Bed Elevation as a Diffusion or Advection Problem**

   Consider the alluvial bed of a river with an elevation above a datum equal to h [m]. If the total bed material load per unit width is q_b [m^3 s^{-1} m^{-1}], derive an equation relating temporal changes in bed elevation to spatial changes in q_b. Now assume that

$$q_b = k\tau^a,$$

where k and a are constants for a given sediment type, and the bed shear stress is given by

$$\tau = \rho g(\eta - h)S,$$

where ρ is the fluid density, g is gravitational acceleration, η is the (undisturbed) height of the water surface above the reference level, and S is the bed slope. Let $H = (\eta - h)$ and derive a first-order PDE that can be solved for H. Use the resulting equation to predict qualitatively the evolution of an initial bar or bedform of positive relief.

Advection and Diffusion (Transport) Problems

Consider a property P that is conserved. It could be mass such as the amount of a dissolved species in a river or the mass of particulate load suspended in a flow, or it could be a vector property of a mudflow like momentum. Further assume that P is passively carried along by the medium in which it exists at the flow speed u; that is, it is advected along. Also assume that within the medium, P moves from one point to another in proportion to its gradient; that is, it diffuses according to a first-order rate law. Finally, assume that there is a source and/or sink for P in the medium such as might occur if P is converting to another chemical species at a known rate. The resulting PDE is called the *transport equation.*

Transport equations combine the concepts of diffusion and advection presented in earlier chapters. As such, they describe an amazing gamut of geological processes. In oceans and rivers, they allow the prediction of salt, heat, dissolved and suspended particles, and even the turbulent kinetic energy of the flow. In lavas and volcanic eruptions, they predict the evolution of heat. In the seabed, transport equations predict the evolving chemistry of the pore waters. As one could imagine, there is a wealth of literature on transport phenomena. Probably the most seminal and still the most accessible for geoscientists is the textbook by Bird, Stewart, and Lightfoot called *Transport Phenomena*. It was first published in 1960 and now is in its 60th (!) edition. In the geosciences, a good general text is Turcotte and Schubert's (1982) *Geodynamics*. For specialists there is Bear and Buchlin's (1991) *Modelling and Applications of Transport Phenomena in Porous Media*, Clark's (2009) *Transport Modeling for Environmental Engineers and Scientists*, and Boudreau's (1996) *Diagenetic Models and Their Interpretation*. Because transport modeling spans so many fields in the earth sciences, we start with a generic model in one dimension. No new concepts are needed to extend the equation to more dimensions.

Translations

Example I: A Generic 1-D Case

Physical Picture

Consider the geometry in figure 7.1, where a conservative substance or property P flows through x–y space at rate u and also diffuses. It also may have a source or sink. We seek $P(x,t)$ where for the purposes of concreteness, P will be a concentration [kg m^{-3}] and x and t are distance [m] and time [s], respectively. Assume that the thickness and depth of the control volume are Y and W, respectively, q_P is a diffusive flux [kg m^{-2} s^{-1}], and the source or sink is S [kg m^{-3} s^{-1}].

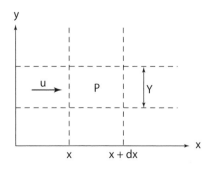

Figure 7.1. Definition sketch for transport of a passive property $P(x,t)$, where u is the advection speed of the medium [m s^{-1}], and Y is a finite thickness [m].

Physical Laws

With one independent variable, one equation is needed. It is important to recognize from the outset that concentrations are not a conservative property (they cannot be added); thus, to apply the conservation laws, we must convert concentration into a conservative property. The obvious choice is conservation of mass, and the obvious control cell is given in figure 7.1.

Restrictive Assumptions

We assume the only processes of mass transport are advection and diffusion, and the geological situation is such that one dimension is sufficient.

Perform the Balance

Thus,

$$\text{TROCM} = \text{MRI} - \text{MRO} + \text{SOURCE/SINK}, \quad (7.1)$$

or

$$\frac{\partial PYW dx}{\partial t} = PuYW - \left[PuYW + \frac{\partial PuYW}{\partial x} dx \right] \\ + q_p W - \left[q_p W + \frac{\partial q_p W}{\partial x} dx \right] + SYW dx. \quad (7.2)$$

The first two terms on the RHS are respectively the net rate of addition of mass of P to the cell due to flow into the face at x and out of the face at $x + dx$. The second two terms represent the net rate of addition of P to the cell due to diffusion, and the last term is the source (if positive) or sink (if negative). As in chapter 3, the diffusive flux is usually approximated as

$$q_p = -D \frac{\partial P}{\partial x}, \quad (7.3)$$

where D is the diffusivity [m^2 s^{-1}]. Under the assumption that Y, W, and D do not vary in time or space, equation 7.3 can be substituted into equation 7.2 and constants taken outside of the differentials to yield

$$\frac{\partial P}{\partial t} + u \frac{\partial P}{\partial x} - D \frac{\partial^2 P}{\partial x^2} - S = 0. \quad (7.4)$$

This is the transport equation with a source or sink for the case where the diffusivity does not vary with x.

Check Units

Units of each term are kg m^{-3} s^{-1}.

Define Interval, Specify Initial and Boundary Conditions

These depend upon the problem of interest of course, but for completeness let the intervals be $x = 0$ to L and $t = 0$ to L/u and the initial condition be $P(x,0) = 0$, and the boundary conditions be $P(0,t) = P_o$ and $P(L,t) = 0$.

To learn more about the behavior of its solutions, it is useful to nondimensionalize equation 7.4 by the following transformations:

$$t^* = ut/L$$
$$x^* = x/L \tag{7.5}$$
$$P^* = P/P_o,$$

where P_o is some standard value of the property P. Substituting these into equation 7.4 yields

$$\frac{\partial P^*}{\partial t^*} + \frac{\partial P^*}{\partial x^*} - \frac{D}{uL}\frac{\partial^2 P^*}{\partial x^{*2}} - \frac{SL}{P_o u} = 0. \tag{7.6}$$

The coefficient of the third term is a dimensionless number representing the relative rates of diffusion to advection. As noted in chapter 5, the ratio of advection to diffusion of a property is called the Peclet number. Thus, this coefficient is the inverse of the Peclet number. What class of equation is this? Applying the classification method developed in chapter 1 reveals that it is parabolic like the diffusion equation of chapter 4. However, note that if the Peclet number is large, then the coefficient of the diffusion term in equation 7.6 is small, and in the limit, equation 7.6 reduces to the advection equation, which in chapter 6 we noted was hyperbolic. It follows, then, that equation 7.6 is either parabolic or hyperbolic depending upon whether the magnitude of the Peclet number is larger or smaller than 1.

At intermediate Peclet numbers, the behavior is demonstrated in figure 7.2. Note that if $S = 0$, the area under

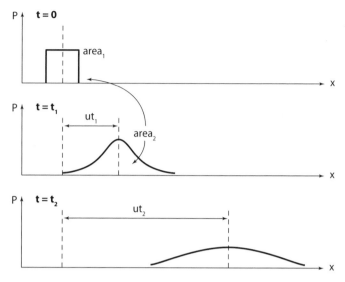

Figure 7.2. Solution to the transport equation when the magnitude of the Peclet number is of $O(10^0)$. Note that the area under the curve is conserved (area$_1$ = area$_2$) if there are no sources or sinks.

the curve is conserved as the property P advects downstream at the speed u and diffuses laterally.

Example II: Transport of Suspended Sediment in a Stream

One-dimensional transport of suspended sediment provides a good geological example of the transport equation. In turbulent open channel flows such as rivers and tidal channels, bed sediment of sufficiently small grain settling velocity is borne aloft by the net upwards-directed turbulent momentum flux where it travels at essentially the free stream velocity. The bulk of the world's sediment travels to the sea in this manner to build deltas and renourish the world's coastlines. Along the way it adsorbs pollutants, sometimes gets trapped behind dams, and chokes irrigation canals. Predicting its behavior as a function of the relevant flow variables is an important objective in sedimentary geology and environmental geosciences. Here we show how one derives a model predicting 1-D suspended load transport. For a more in-depth treatment, see Gyr and Hoyer (2006) and Clark (2009).

Physical Picture

Consider the physical picture in figure 7.3 where a river of depth H [m] flowing at the speed u [m s^{-1}] in the x direction carries suspended sediment particles at concentration $C(x,t)$ [kg m^{-3}]. We want to know how C varies with distance down-river and time. Assume for simplicity that variations in the cross-stream direction are not of interest, and the particles all have the same constant terminal settling velocity w [m s^{-1}].

Physical Laws

The function $C(x,t)$ could be defined using the law of conservation of mass if we knew what processes were responsible for carrying particulates into and out of the control volume sketched in figure 7.3. Their advection into the box at the x-face by the flow itself would seem to be

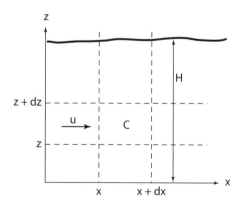

Figure 7.3. Vertical longitudinal slice along a river channel showing control cell for suspended sediment problem. x is downstream, z is normal to the bed, and thickness of cell in the y direction is T [m].

important. They could be convected through the z-face, too, but by assuming that the flow is solely in the x direction, this can be ignored. A flux in the z direction that can't be ignored, however, is that due to grain settling. Grains larger than a few micrometers settle out of the flow due to the pull of gravity. That this isn't the end of the story is readily apparent, because rivers maintain their turbidity in the face of this settling flux. It was Sir Reginald A. Bagnold, the British sedimentologist and engineer, who first argued that grains were kept aloft in geophysical flows by a net upwards-directed momentum flux in the fluid due to the anisotropy of turbulence. The simplest characterization of this process is to liken it to Brownian motion, in which case it is reasonable to suppose that the sediment flows from points of higher concentration to points of lower concentration; that is, it follows a Fickian first-order rate law,

$$q_z = -D\frac{\partial C}{\partial z}, \tag{7.7}$$

where q_z is the mass flux of sediment per unit area, and D is the diffusivity [$m^2\ s^{-1}$], also called the kinematic eddy diffusivity of the suspended sediment. D is not a constant but depends upon the intensity of turbulence at each point in the flow and therefore is a function of z, x, and t. In summary, we assume that sediment can be advected into the cell in the x direction and diffused into and settle out of the cell in the z direction, but sediment diffusion in the x direction is much less important than sediment advection.

Restrictive Assumptions

Assume a stream of constant width and a single grain size of interest.

Perform the Balance

Putting all of this into words,

TROCM = MRI due to advection
    − MRO due to advection + MRI due to
    diffusion and settling − MRO due to
    diffusion and settling, $\hspace{2em}$ (7.8)

and in symbols,

$$\frac{\partial CTdxdz}{\partial t} = CuTdz - \left[CuTdz + \frac{\partial CuTdz}{\partial x}dx \right]$$

$$+ q_zTdx - \left[q_zTdx + \frac{\partial q_zTdx}{\partial z}dz \right] \qquad (7.9)$$

$$- CwdxT - \left[-CwdxT - \frac{\partial CwdxT}{\partial z}dz \right],$$

where T is the thickness of the cell in the y direction. The first grouping of terms on the RHS is the net rate of mass addition due to advection in the alongstream direction, the second set is the net rate of addition of sediment to the cell from turbulent diffusion in the vertical, and the third set is the net rate of addition due to particle settling. Notice that it is negative because the z axis is positive upwards. Simplifying and canceling like terms yields

$$\frac{\partial C}{\partial t} + \frac{\partial Cu}{\partial x} + \frac{\partial q_z}{\partial z} - w\frac{\partial C}{\partial z} = 0. \qquad (7.10)$$

Now we can substitute in the definition of q_z (equation 7.7) and use the product rule to simplify:

$$\frac{\partial C}{\partial t} + \frac{\partial Cu}{\partial x} - D\frac{\partial^2 C}{\partial z^2} - \frac{\partial D}{\partial z}\frac{\partial C}{\partial z} - w\frac{\partial C}{\partial z} = 0. \qquad (7.11)$$

Check Units

Units of each term are kg m^{-3} s^{-1}.

To solve equation 7.11, one needs to know how D varies in space and time. By the Reynolds analogy, the sediment diffusivity usually is taken to be equal to the fluid turbulent diffusivity, often called the eddy viscosity. As explained in chapter 8, the vertical fluid eddy viscosity in an open channel flow is approximated by

$$D = \kappa\sqrt{\frac{\tau_o}{\rho}}z\left(1 - \frac{z}{H}\right), \qquad (7.12)$$

where κ is equal to 0.4, τ_o is the bed shear stress, and ρ is the fluid density.

With appropriate ICs and BCs, equation 7.11 and equation 7.12 could be solved for suspended sediment concentration as a function of time, space, grain settling velocity, and flow velocity. But to make the point that this is a transport equation in disguise, now further assume that the flow is steady (so the first term equals zero) and uniform (so that the u can be moved outside of the differential in the second term. Differentiating D with respect to z then yields

$$\frac{\partial C}{\partial x} - \frac{D}{u}\frac{\partial^2 C}{\partial z^2} - A\frac{\partial C}{\partial z} = 0$$

where

$$A = \frac{\kappa\sqrt{\frac{\tau_o}{\rho}}\left(1 - \frac{2z}{H}\right) + w}{u}. \tag{7.13}$$

This is a form of the transport equation, although here it is the gradient of concentration along the stream rather than the gradient in time that changes as a result of advection and diffusion of the particles in the vertical. Also, the coefficients are not constant but functions of z.

Define Interval, Specify Initial and Boundary Conditions

Let the intervals be $x = 0$ to L and $z = 0$ to H. There is no initial condition because this is a steady-state problem. Let the boundary conditions be $C(0,z) = J$, $C(x,0) = C_o$, and $C(x,H) = 0$. This completes the derivation.

Example III: Sedimentary Diagenesis: Influence of Burrows

The pioneering work of Robert Berner of Yale in the 1970s advanced the field of low-temperature geochemistry into a quantitative science. His book *Early Diagenesis* (Berner, 1980) provided the conceptual and mathematical basis to extract rates of diagenetic processes from profiles of concentration of important aqueous species in sedimentary porewaters. Robert Aller, a Ph.D. student at Yale during this time, advanced the sophistication of Berner's modeling by considering the two-dimensional implications of burrowing by organisms. Some organisms construct

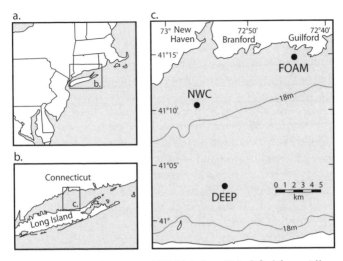

Figure 7.4. Location map of FOAM sites. [Modified from Aller, R. C. (1980). Diagenetic processes near the sediment-water interface of Long Island Sounds: I. Decomposition and nutrient geochemistry (S, N, P). *Advances in Geophysics* 22:237–350.]

"permanent" vertical burrows through which they pump water from the overlying sediment, essentially carrying the sediment-water into the sediment. Berner and his students' quantitative approach provided a clear explanation for an otherwise perplexing observation in his backyard study area, Long Island Sound; porewater profiles there seemed to suggest that sulfate reduction rates, expressed as sulfate depletion in sediments, increased offshore, from a shallow-water site (FOAM; fig. 7.4) to the deeper-water sites NWC and DEEP. However, radiotracer studies indicated that sulfate reduction rates in fact were similar at the three sites. Why would the porewater profiles be so different?

Here we explore these concepts to illustrate how models of this sort are constructed. We also take this opportunity to demonstrate how other coordinates systems—in this case a cylindrical coordinate system—may be used to good advantage when the physical system requires it.

Physical Picture

Consider the case in which a sediment is undergoing the process of microbial sulfate reduction, bioturbation, permanent burrow construction, and sedimentation. The observed profiles in sediments (e.g., depletion of oxygen and sulfate and increase in hydrogen sulfide concentrations) need to be interpreted in terms of both transport and reaction, with transport by both molecular diffusion and a diffusive-like process as well as porewater advection driven by compaction and sedimentation. The burrows can be considered to be vertical and cylindrical, with radius r_o. The sediment is otherwise homogeneous, so that the problem can be considered as a radial and vertical (2-D) diffusion/advection/reaction problem. The burrows are sufficiently far apart that the influence of adjacent burrows can be neglected. Thus, we can consider this a two-dimensional problem with a vertical axis (x) centered on the burrow extending from the sediment–water interface positive downward, and a radial axis perpendicular to this (r). The control volume is a cylindrical shell extending from radius r to $r + dr$ and from height x to $x + dx$ (fig. 7.5).

Physical Laws

Solutes are transported through the connected sediment porewater system by molecular diffusion, but bioturbation typically is so much more effective that we can ignore molecular diffusion. But the formulation is the same: we assume diffusion by bioturbation follows Fick's first law,

$$q = -D_B(s)\frac{\partial C}{\partial s}, \tag{7.14}$$

where q is the flux of solute per unit area in either the x or r direction [mol s^{-1} m^{-2}], $D_B(s)$ is the bioturbation diffusivity [m^2 s^{-1}], C is the concentration of solute [mol m^{-3}], and s is the axis in question (either x or r) [m]. The advective flux of solute with sedimentation is simply the product of the sedimentation rate times the concentration (with the restrictive assumptions below) [mol m^{-2} s^{-1}].

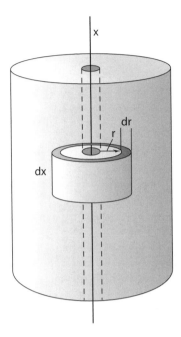

Figure 7.5. Conceptual diagram of the diagenetic environment surrounding a burrow.

Restrictive Assumptions

For simplification, we will assume that porosity ϕ, sedimentation rate ω [m s^{-1}], and sulfate reduction rate R [mol m^{-3} s^{-1}] are constant with depth and in time, and there is no groundwater flow.

Perform the Balance

In this problem, the volume of the computational cell as shown in figure 7.5 is a shell extending from r to $r + dr$ and from x to $x + dx$, with volume $\pi \, dx \, [(r + dr)^2 - r^2]$. Note that as dr is considered to be very small, dr^2 is much smaller and can be ignored. Thus, upon expanding $(r + dr)^2$, the control volume becomes $2\pi r dr dx$. The vertical diffusive flux is through a donut-shaped area of $\pi \, [(r + dr)^2 - r^2]$, which simplifies to $\pi(2rdr)$. Now write the conservation of mass law first in words,

Time rate of change of moles in the cell
= Mole rate in − Mole rate out
+ Sources − Sinks, (7.15)

then in symbols:

$$\frac{\partial C\phi 2\pi r dr dx}{\partial t} = q_x \phi 2\pi r dr - \left[q_x \phi 2\pi r dr + \frac{\partial q_x \phi 2\pi r dr}{\partial x} dx \right]$$

$$+ q_r \phi 2\pi r dx - \left[q_r \phi 2\pi (r + dr) dx + \frac{\partial q_r \phi 2\pi (r + dr) dx}{\partial r} dr \right]$$

$$+ \omega C \phi 2\pi r dr - \left[\omega C \phi 2\pi r dr + \frac{\partial \omega C \phi 2\pi r dr}{\partial x} dx \right]$$

$$- R(\phi 2\pi r dr dx). \quad (7.16)$$

The porosity enters into the flux terms on the RHS of equation 7.16 because we are assuming that the cross-sectional area of pore fluid through which the solute can diffuse, relative to that of solid sediment, is equivalent to the porosity. Distributing $(r + dr)$ in the term

$$\frac{\partial q_r \phi 2\pi (r + dr) dx dr}{\partial r}$$

gives

$$\frac{\partial q_r \phi 2\pi r dx dr}{\partial r} + \frac{\partial q_r \phi 2\pi dr^2 dx}{\partial r},$$

and as dr^2 is very small, the second term can be ignored. Then, canceling terms, dividing through by $2\pi\phi dr dr dx$, as these are not functions of time or space, and substituting in equation 7.14 yields

$$\frac{\partial C}{\partial t} = \frac{\partial}{\partial x} D_B \frac{\partial C}{\partial x} + \frac{D_B}{r} \frac{\partial C}{\partial r} + \frac{\partial}{\partial r} D_B \frac{\partial C}{\partial r} - \omega \frac{\partial C}{\partial r} - R. \quad (7.17)$$

For the special case where D_B is homogeneous,

$$\frac{\partial C}{\partial t} - D_B \left(\frac{\partial^2 C}{\partial x^2} + \frac{\partial^2 C}{\partial r^2} \right) - \left(\frac{D_B}{r} - w \right) \frac{\partial C}{\partial r} + R = 0. \quad (7.18)$$

Check Units
Units are moles per cubic meter.

Define Interval, Specify Initial and Boundary Conditions

Typical intervals could be $r = 0$ to ∞, $z = 0$ to x_{max}, and $t = 0$ to t_{max} with an initial condition of constant concentration C_o everywhere. Typical boundary conditions might be $C(r = 0, t) = C_{seawater}$, and $C(r = \infty, t) = C_o$.

This is clearly a transport equation, with both diffusive (vertical and radial) and advective components. The advection speed is the difference between the burrowing velocity and the sedimentation rate. When these are equal, the system becomes completely diffusive.

Finite Difference Solutions to the Transport Equation

In the 1-D transport equation written in dimensional form (equation 7.4), the time rate of change of a conservative property, P, consists of the sum of an advection and a diffusion term (assuming there is no source or sink for P):

$$\frac{\partial P}{\partial t} + u\frac{\partial P}{\partial x} - D\frac{\partial^2 P}{\partial x^2} = 0, \tag{7.19}$$

where u is the advection speed and D is the diffusivity. Given earlier discussions on finite difference schemes, you might assume that an explicit FTCS scheme would suffice here in which both the advection and diffusion terms are discretized by central differences:

$$\frac{P_i^{n+1} - P_i^n}{\Delta t} = -u\frac{P_{i+1}^n - P_{i-1}^n}{2\Delta x} + D\frac{P_{i-1}^n - 2P_i^n + P_{i+1}^n}{\Delta x^2}. \tag{7.20}$$

Unfortunately, for equation 7.20 to be accurate, it must be true that

$$\Delta t \leq \frac{\Delta x^2}{2D} \quad \text{and} \quad \Delta t \leq \frac{2D}{u^2}, \tag{7.21}$$

and these conditions are often impossible to honor, especially in geological problems. Furthermore, if the problem is a pure advection problem, $D = 0$. Then the criteria are impossible to honor, indicating that the scheme is unconditionally unstable under those conditions. The difficulty arises because the transport equation takes on properties of either a parabolic or hyperbolic equation, depending upon the magnitude of the Peclet number. Clearly, we need a more robust scheme.

Many schemes have been proposed for equation 7.19 because the transport equation appears so often in science and engineering studies, but each has its drawbacks (e.g., see Gajdos and Mandelkern, 1998). Particular inaccuracies include numerical diffusion, phase errors, and introduction of higher-frequency waves—the "artificial wiggles" of numerous authors. Here we present what to us is the simplest of the reasonably robust schemes—the explicit QUICK and QUICKEST schemes of Leonard (1990). QUICK can be used for transport problems with steady or quasi-steady flows, whereas QUICKEST should be used if the flow is unsteady. But it should be noted that if the property in question must remain positive (such as salt concentration), then even QUICKEST may not be sufficient for cases with sharp fronts. In these cases, more complicated flux-limiting schemes must be used.

QUICK Scheme

First we derive the QUICK scheme. Consider the finite difference cells in figure 7.6, where the values of P are to be found at the cell centers. In the second term from the left in equation 7.20, the convective flux into the cell through the left wall (at l in fig. 7.6) minus the flux out at the right wall (r in fig. 7.6) is estimated by values at $i - 1$ and $i + 1$.

Intuitively, it would seem that a more accurate estimate would use the difference between $P|_{x=r}$ and $P|_{x=l}$; that is,

Figure 7.6. Finite difference line for 1-D transport equation. Vertical lines represent finite difference cell walls. l = left; r = right.

$$\frac{P_i^{n+1} - P_i^n}{\Delta t} = -u\frac{P_r^n - P_l^n}{2\Delta x} + D\frac{P_{i-1}^n - 2P_i^n + P_{i+1}^n}{\Delta x^2}. \quad (7.22)$$

Furthermore, in estimating the magnitudes of P at r and l, it would be good to fit a curve through the function $P(x)$ rather than a straight line. In the QUICK method, P_l and P_r are estimated by fitting a quadratic equation to the values of P in the vicinity of node i. Starting with P_r first, let

$$P_r = Ax^2 + Bx + C, \quad (7.23)$$

where A, B, and C are coefficients to be determined. To find P_r take $x = 0$ at the right cell wall. With $x = 0$, equation 7.23 reduces to $P_r = C$. How do we determine C? We write three equations and solve for A, B, and C simultaneously. The value at P_{i+1} which is $\Delta x/2$ away from position r is written as the quadratic function:

$$P_{i+1} = A\left(\frac{\Delta x}{2}\right)^2 + B\left(\frac{\Delta x}{2}\right) + C. \quad (7.24)$$

Likewise, P_i becomes

$$P_i = A\left(\frac{-\Delta x}{2}\right)^2 + B\left(\frac{-\Delta x}{2}\right) + C. \quad (7.25)$$

and P_{i-1}:

$$P_{i-1} = A\left(\frac{-3\Delta x}{2}\right)^2 + B\left(\frac{-3\Delta x}{2}\right) + C \quad (7.26)$$

Equation 7.24 to equation 7.26 constitute a set of three equations that can be solved for the coefficients A, B, and C in terms of the values of P at the nodes and Δx. Substituting the values of A, B, and C in equation 7.23 yields

$$P_r = \frac{3}{4}P_i - \frac{1}{8}P_{i-1} + \frac{3}{8}P_{i+1}. \quad (7.27)$$

By a similar approach, P_l can be defined as

$$P_l = \frac{3}{8}P_i + \frac{3}{4}P_{i-1} - \frac{1}{8}P_{i-2}. \quad (7.28)$$

Substituting equation 7.27 and equation 7.28 into equation 7.22 and solving for P_i^{n+1} yields:

$$P_i^{n+1} = P_i^n - c\left(\frac{1}{8}P_{i-2}^n - \frac{7}{8}P_{i-1}^n + \frac{3}{8}P_i^n + \frac{3}{8}P_{i+1}^n\right)$$
$$+ \alpha\left(P_{i-1}^n - 2P_i^n + P_{i+1}^n\right), \tag{7.29}$$

where the Courant number is given by:

$$c = \frac{u\Delta t}{\Delta x},$$

and the diffusion parameter is

$$\alpha = \frac{D\Delta t}{\Delta x^2}.$$

This is the QUICK scheme. It is highly accurate and stable if

$$\alpha + \frac{c}{4} \le \frac{1}{2} \quad \text{and} \quad c^2 \le 2\alpha.$$

QUICKEST Scheme

The QUICKEST scheme is somewhat more complicated because it uses not only gradients but also curvatures to estimate the differentials. For constant c and Δx, the QUICKEST finite difference equation is

$$P_i^{n+1} = P_i^n - c\left[\left[\frac{1}{2}\left(P_i^n + P_{i+1}^n\right) - \frac{\Delta x}{2}c\,\text{GRAD}_r\right.\right.$$
$$-\frac{\Delta x^2}{6}(1 - c^2 - 3\alpha)\text{CURV}_r\right) - \left(\frac{1}{2}\left(P_{i-1}^n + P_i^n\right)\right.$$
$$\left.\left.-\frac{\Delta x}{2}c\,\text{GRAD}_l - \frac{\Delta x^2}{6}(1 - c^2 - 3\alpha)\text{CURV}_l\right)\right] \tag{7.30}$$
$$+ \alpha\left[\left(\Delta x\,\text{GRAD}_r - \frac{\Delta x^2}{2}c\,\text{CURV}_r\right)\right.$$
$$\left.-\left(\Delta x\,\text{GRAD}_l - \frac{\Delta x^2}{2}c\,\text{CURV}_l\right)\right],$$

where

$$\text{GRAD}_l = \left(P_i^n - P_{i-1}^n\right)/\Delta x$$
$$\text{CURV}_l = \left(P_{i-2}^n + P_i^n - 2P_{i-1}^n\right)/\Delta x^2$$
$$P_i^n = \frac{1}{2}\left(P_{i-1}^n + P_i^n\right) - \frac{\Delta x^2}{8}\text{CURV}_l$$

and

$$\text{GRAD}_r = \left(P^n_{i+1} - P^n_i \right)/\Delta x$$

$$\text{CURV}_r = \left(P^n_{i-1} + P^n_{i+1} - 2P^n_i \right)/\Delta x^2$$

$$P^n_r = \frac{1}{2}\left(P^n_i + P^n_{i+1} \right) - \frac{\Delta x^2}{8}\text{CURV}_r.$$

Equation 7.30 has a stability criterion that is a complex function of both c and α. For the complete stability region, see Leonard (1990).

How good is the QUICKEST scheme? Figure 7.7 shows a solution computed by QUICKEST and an explicit upwind scheme for a generic initial condition with advection and diffusion. When compared with the exact solution, QUICKEST does not show the numerical diffusion of the upwind scheme.

It should be noted that QUICK and QUICKEST allow unnatural extrema, including negative values that may be unacceptable (e.g., if one is modeling biogeochemical processes where negative concentrations are to be avoided). In such cases, higher-order, monotonic "flux limited" schemes should be employed (e.g., Thuburn, 1996).

Summary

The transport equation seems to be everywhere, and that is because it describes one of the most common combination of processes in geology—advection plus diffusion of a conservative property. The behavior of the solution depends upon the relative magnitudes of the advection and diffusion terms, and thus on the Peclet number.

Modeling Exercises

1. Using Salt to Estimate Discharge

   Consider a river of constant rectangular cross section $A = 200$ m^2 and steady, uniform discharge $Q = 100$ m^3 s^{-1}. At a point $x = 0$ m, and for 1,000 seconds ($0 < t < 1{,}000$ s), 100 kg s^{-1} of salt is poured into the stream, instantly dissolving and mixing top-to-bottom and across (but not along)

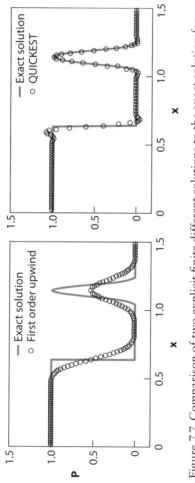

Figure 7.7. Comparison of two explicit finite different solutions to the exact solution for a generic transport problem. Note that the upwind solution contains artificial diffusion, whereas QUICKEST does not. [Modified from Le Roy, S. (2006). Numerical methods in the simulation of charge transport in solid dielectrics. *IEEE Transactions on Dielectrics and Electrical Insulation* 13(2):239–246.]

the channel. Assume that the salt moves both by diffusion and advection in the direction of flow.

a. Derive a well-posed model from first principles and following the proper steps in model building to describe the concentration of salt along the length of the stream.

b. Using QUICKEST, calculate the "breakthrough" curve for salt concentration 1,000 m downstream for the case in which the salt diffusivity in the turbulent stream is 1×10^{-4} m^2 s^{-1}.

2. **Bubbles in Ice**

As new snow is deposited on top of old snow, air is trapped. As the snow layer is buried by new snow at deposition rate w, the snow becomes firm, then ice, and bubbles of the air become encased in the ice. Later, glaciologists can come along, core the ice, and extract the bubbles. Measurements of gas composition can be made, and ancient atmospheric concentrations can be determined. However, gas can diffuse through the ice according to Fick's law, with a diffusion coefficient D [m^2 s^{-1}].

a. Derive a properly posed mathematical model for the variation of CO_2 concentration C [mol of CO_2 m^{-3} of ice] as a function of time and depth including diffusion and snow deposition w [m y^{-1}], assuming that compaction is negligible. Hint: Assign the snow–air interface as the upper boundary ($x = 0$), which moves upward at the snowfall rate w. Your upper boundary condition is a function of time.

b. Calculate the burial history of the anthropogenic CO_2 perturbation simplifying the curve as a baseline of 280 ppm for the last several thousand years and a ramp from this value at 1,900 to a stabilized 450 ppm by the year 2100, buried at a constant rate $w = 0.1$ m y^{-1}. Show the profile at 500 and 5,000 years into the future. Assume $D = 1 \times 10^{-7}$ cm^{-2} s^{-1}, and that CO_2 solubility in ice is 2×10^{-6} mol cm^{-3} ppm^{-1}.

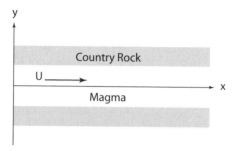

Figure 7.8. Schematic diagram for modeling exercise 3. Country rock denoted by gray; $y = 0$ defines center of magma channel along which magma flows in the positive x direction at a velocity U [m s^{-1}].

3. **Cooling of a 1-D Magma Sill**

Figure 7.8 depicts a sill of molten magma injected between two layers of thermally isotropic rock. The magma is flowing in the positive x direction at velocity U [m s^{-1}]. As it flows, it warms the country rock (i.e., heat is extracted from the magma) following Newton's law of cooling:

$$q = h(T - T_c),$$

where q = heat flux out of the magma [calories per second per unit area]; h = heat exchange coefficient [calories per second per m^2 per unit of temperature difference]; T = temperature of the magma at location x and time t [°C]; and T_c = temperature of the country rock [°C], assumed to be constant in time and space. Other useful thermal properties of the materials are: diffusivity, D [m^2 s^{-1}]; thermal conductivity, k [calories per length per time per °C]; and thermal capacity, c [calories per unit mass per °C].

Derive the 1-D mathematical model describing the temperature of the magma as a function of time and space (x direction). Assume plug flow of the magma (i.e., assume no variation in magma temperature or velocity in the y direction).

Transport Problems with a Twist: The Transport of Momentum

In chapter 7, we explored the transport of a property by advection and diffusion. Whether the property was a mass of suspended sediment, dissolved ions in a stream, or heat in a lava flow, the resulting PDEs possessed the same form, and consequently the solutions behaved similarly. A signal entering the domain of interest from a boundary, or a function describing an initial condition, was translated across the domain while diffusing away. The amount of translation relative to diffusion could be qualitatively predicted if one knew the ratio of the advection speed to diffusivity (say in the form of a Peclet number).

This chapter considers a special case of the transport equation where the property in question is momentum. As we shall see, when the property being transported is momentum, the convective term in the transport equation becomes nonlinear. This leads to even richer behavior of the solutions. Equations of this type provide the basis for computing the velocities and discharges of all geophysical fluid flows. Examples include Euler's, St. Venant's, Navier–Stokes', and Reynolds' equations. But before delving into these complete equations of motion, it is illuminating to consider a special form without sources of momentum such as pressure forces, and sinks such as friction. This simplified form was first explored by Johannes Martinus Burgers (1895–1981), after whom it is named. For a more in-depth discussion of the Burgers' equation and its behavior, see Tannehill et al. (1997).

Translations

Example I: One-Dimensional Transport of Momentum in a Newtonian Fluid (Burgers' Equation)

Physical Picture

Consider the domain in figure 8.1 in which a control volume or cell of dimensions dx by dy by dz sits in a nonturbulent *Newtonian fluid*. A Newtonian fluid is one in which the shear stress between adjacent planes of fluid is linearly proportional to the velocity gradient between the planes; that is to say, it obeys Newton's law of viscosity. We want to know how the momentum of the fluid varies with distance along the x axis [m] and with time, t [s] (i.e., it is a 1-D problem). Without too much loss of generality, we can consider the property of interest to be momentum *per unit mass*, making the dependent variable just velocity $u(x,t)$ [m s^{-1}].

Physical Laws

An obvious place to start is conservation of momentum. For the moment we ignore any external forces, yielding

The time rate of change of x-directed momentum
    in the control volume (TROCMOM$_x$) = The
    x-momentum rate into the volume (MOMRI$_x$)
    − The x-momentum rate out (MOMRO$_x$).   (8.1)

For this nonturbulent Newtonian fluid there are two processes whereby x-directed momentum can enter the control volume. The first is advection of momentum. Fluid flowing through the yz face at location x carries momentum into the cell because there is a mass flux through the face and that mass possesses a velocity u (remember that momentum is defined as m *times u*). The second process is the diffusion of momentum through the face at x. As discussed in chapter 4, the way to think about this diffusion process is to consider some fluid molecules of unit mass just upstream of, and some just downstream of, the face. Their x-directed momentum

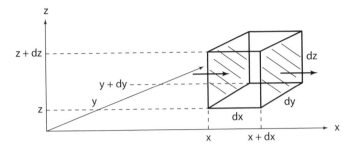

Figure 8.1. Definition sketch for derivation of Burgers' equation.

in each case is given by their velocity, u. By Brownian motion, these molecules will be interchanging through the face at a certain rate. If their velocities are not the same, then in unit time there will be an addition or loss of momentum from the control volume even though the mass fluxes from Brownian motion sum to zero. The greater the difference in velocities across the face, the greater will be the addition or loss of momentum. Therefore, it seems sensible to assume that the diffusional flux of momentum across the face at x is proportional to the velocity gradient, or

$$q_x = -\mu \frac{\partial u}{\partial x}, \qquad (8.2)$$

where q_x is the diffusive momentum flux [kg m^{-1} s^{-2}], μ is the proportionality constant [kg m^{-1} s^{-1}] (which is the molecular viscosity), and the negative sign ensures that momentum flows in the correct direction relative to the velocity gradient. This first-order rate law was presented in chapter 1 and is one form of Newton's law of viscosity. Notice that in this case, it is a normal rather than a tangential force. The same ideas pertain to the faces perpendicular to the y and z axes, but momentum fluxes through these faces can be neglected because this is a 1-D problem.

Restrictive Assumptions

We have assumed no sources or sinks of momentum for a Newtonian fluid.

Perform the Balance

Substituting variables for words in equation 8.1 and using Taylor series as before to define the momentum flux out of the volume at the $x + dx$ face yields

$$\frac{\partial \rho dx dy dz u}{\partial t} = \rho dy dz uu - \left(\rho dy dz uu + \frac{\partial \rho dy dz uu}{\partial x} \right) dx$$
$$+ q_x dy dz - \left(q_x dy dz + \frac{\partial q_x dy dz}{\partial x} dx \right), \tag{8.3}$$

where ρ is the fluid density [kg m^{-3}]. The LHS is the time rate of change of momentum in the control cell. It is given by the fluid density times the volume of the cell times the velocity of the fluid in the cell. The first term on the RHS is the advective momentum flux into the cell through the face at x. It can be thought of as a mass flux through the face, $\rho dy dz u$, which carries an amount of momentum in it equal to u. The next term on the RHS is the momentum flux out of the cell at $x + dx$ obtained by Taylor series. The third term on the RHS is the momentum flux into the cell at x due to molecular diffusion. It consists of the diffusive momentum flux per unit area times the area of the face. Canceling terms and dividing through by the volume of the cell yields

$$\frac{\partial \rho u}{\partial t} = -\frac{\partial \rho u^2}{\partial x} - \frac{\partial q_x}{\partial x}. \tag{8.4}$$

This is the conservative form of the momentum equation, so called because it represents both conservation of momentum and fluid mass. Equation 8.4 can be simplified by considering the statement of conservation of fluid mass for this case. Stated in words:

> The time rate of change of fluid mass in the volume
> = The mass rate in − The mass rate out. (8.5)

Substituting symbols for words and following the procedures in earlier chapters yields

$$\frac{\partial \rho}{\partial t} + \frac{\partial \rho u}{\partial x} = 0. \tag{8.6}$$

This is a statement of conservation of mass of the fluid. Now continue simplifying equation 8.4 using the product rule,

$$\rho \frac{\partial u}{\partial t} + u \frac{\partial \rho}{\partial t} = -\rho u \frac{\partial u}{\partial x} - u \frac{\partial \rho u}{\partial x} - \frac{\partial q_x}{\partial x}, \tag{8.7}$$

and note that the second and fourth terms can be written as

$$u \left(\frac{\partial \rho}{\partial t} + \frac{\partial \rho u}{\partial x} \right),$$

which by conservation of mass (equation 8.6) is identically zero. Therefore, if the fluid is incompressible such that ρ does not vary in time or space, equation 8.7 becomes

$$\frac{\partial u}{\partial t} = -u \frac{\partial u}{\partial x} - \frac{1}{\rho} \frac{\partial q_x}{\partial x}. \tag{8.8}$$

Substituting equation 8.2 into equation 8.8 yields

$$\frac{\partial u}{\partial t} + u \frac{\partial u}{\partial x} - \nu \frac{\partial^2 u}{\partial x^2} = 0, \tag{8.9}$$

where ν is the kinematic viscosity (i.e., μ/ρ).

Check Units

The units are correctly balanced because each term has units of acceleration.

Define Interval, Specify Initial and Boundary Conditions

Typical intervals would be $x = 0$ to L and $t = 0$ to ∞. Initial conditions can be any continuous function of $u(x,t)$. Two BCs are needed, such as $u(0,t) = u_o$ and $u(L,t) = 0$.

Equation 8.9 is called Burgers' equation. It has the form of a transport equation, but in this case it is momentum per unit mass, or u that is being transported by advection and diffusion. What is notable about equation 8.9 is the fact that the second term is nonlinear. The rate that momentum is advected at a location x depends upon the magnitude of momentum at that location. As noted in chapter 6 in the discussion on lahars, this has profound consequences (e.g., fig. 8.2). Because points of higher velocity advect faster, a shock front forms, and the solution can even become multivalued. Whether a shock manifests itself depends upon the magnitude of the viscosity term relative to the advection speed, as viscosity acts to damp the strong velocity gradients with time and inhibit shock

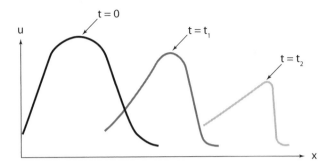

Figure 8.2. Solutions to Burgers' equation for three different times. Note that the initially symmetrical distribution of velocities translates and steepens because the points with higher velocity travel faster. The solution decays with time because of diffusion.

formation. A second point about the nonlinear advective term is that it can create new wave modes. This is evident when a sine wave progressively transforms into a steep front. The front cannot be represented by a single wave form, but represents a sum of waves with shorter wavelengths. These shorter waves arise through the advective term. For example, assuming that $u = u_0 \sin(kx)$ and substituting it into the advection term yields

$$u \frac{\partial u}{\partial x} = u_0^2 k \sin(kx) \cos(kx)$$
$$= u_0^2 \frac{k}{2} \sin(2kx). \tag{8.10}$$

A new wave has arisen with wave number $2k$, or half the wavelength of the initial wave.

Equation 8.9 with momentum sources and sinks added is one of the most commonly occurring PDEs in all of earth science. It is found wherever there are unsteady, nonuniform fluid flows. In subsequent chapters we explore such flows in a river channel and a coastal ocean.

An Analytic Solution to Burgers' Equation

Nonlinear, second-order equations such as Burgers' equation are analytically solvable for simple ICs and BCs, although the solutions can be complex. We provide one example here to act as a standard for numerical solution schemes later. First, nondimensionalize equation 8.9 using the definitions:

$x/L \rightarrow x$

$uL/v \rightarrow u$

$vt/L^2 \rightarrow t$

such that equation 8.9 becomes

$$\frac{\partial u}{\partial t} + \frac{\partial E}{\partial x} - \frac{\partial^2 u}{\partial x^2} = 0, \tag{8.11}$$

where

$$E = \frac{u^2}{2}.$$

One also often sees equation 8.11 written in terms of the Jacobian as

$$\frac{\partial u}{\partial t} + A\frac{\partial u}{\partial x} - \frac{\partial^2 u}{\partial x^2} = 0$$

where

$$A = \frac{\partial E}{\partial u}. \tag{8.12}$$

We will make use of this Jacobian form in later chapters.

Define Initial and Boundary Conditions

A solution to equation 8.11 defined for all x in the interval [−5,5] for $t > 0$ is

$$u(x,t) = \frac{x/t}{1 + \sqrt{t}\,\exp[x^2/4t]} \tag{8.13}$$

(analytic solution from Hoffman et al., 2001).

Finite Difference Scheme for Burgers' Equation

The nonlinearity of equation 8.11 would seem to provide an additional hurdle in the search for a stable, accurate, finite difference scheme. For example, consider a solution that has reached the state given in figure 8.3 where a shock front is developing. Approximating the gradients and curvatures near the shock becomes very difficult. Nevertheless, if the convective term is not too large relative to the diffusive term, even simple explicit schemes can provide a solution. For example, one can write an explicit FTCS scheme as

$$\frac{u_i^{n+1} - u_i^n}{\Delta t} + \frac{E_{i+1}^n - E_{i-1}^n}{2\Delta x} = \frac{u_{i+1}^n - 2u_i^n + u_{i-1}^n}{(\Delta x)^2}, \quad (8.14)$$

where the advective and diffusive terms are centered in space. In chapter 6 it was shown that the explicit FTCS scheme for the advection equation was unconditionally unstable. Surprisingly, when this scheme is applied to the transport equation, it becomes stable. The viscous term acts to damp out the perturbations that grew unbounded in the linear advection case. The stability requirements

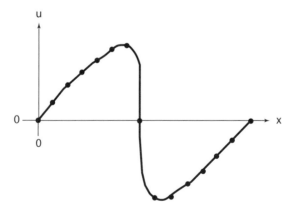

Figure 8.3. Hypothetical solution to Burgers' equation for $u(x,t)$ illustrating difficulties in estimating gradient and curvature terms.

are fairly restrictive, however. If a diffusion number is defined as $s = \nu \Delta t / \Delta x^2$ and a cell Reynolds number as $\text{Re}_c = u \Delta x / \nu$, the stability requirements for the FTCS scheme (equation 8.14) are $s \leq 1/2$ and $\text{Re}_c \leq 2/C$, where C is the Courant number defined earlier as $u \Delta t / \Delta x$. The severity of these restrictions means that one is not free to choose space and time steps most appropriate for the system being studied, but rather must choose them to ensure stability. This suggests that an implicit scheme might give more flexibility while still being accurate.

One such scheme is a FTCS implicit interpretation of equation 8.12 in which the Jacobian is lagged in time:

$$\frac{u_i^{n+1} - u_i^n}{\Delta t} + u_i^n \frac{u_{i+1}^{n+1} - u_{i-1}^{n+1}}{2\Delta x} = \frac{u_{i+1}^{n+1} - 2u_i^{n+1} + u_{i-1}^{n+1}}{(\Delta x)^2}, \qquad (8.15)$$

Equation 8.15 can be rearranged into a tridiagonal system of equations using the formula

$$a_i u_{i-1}^{n+1} + b_i u_i^{n+1} + c_i u_{i+1}^{n+1} = D_i, \qquad (8.16)$$

where

$$a_i = -\frac{\Delta t}{(\Delta x)^2} - u_i^n \frac{\Delta t}{2\Delta x}$$

$$b_i = 1 + 2\frac{\Delta t}{(\Delta x)^2}$$

$$c_i = -\frac{\Delta t}{(\Delta x)^2} + u_i^n \frac{\Delta t}{2\Delta x}$$

$$D_i = u_i^n.$$

A third scheme that is widely used in computational fluid dynamics, the MacCormack method, is neither explicit nor implicit, but of the predictor–corrector type. It calculates the new values of u in two steps. In the predictor step, a "provisional" value of u at time level $n + 1$ (denoted by an overline) is estimated by a forward difference of the advection term

$$u_i^{\overline{n+1}} = u_i^n - \frac{\Delta t}{\Delta x}\left(F_{i+1}^n - F_i^n\right) + r\left(u_{i+1}^n - 2u_i^n + u_{i-1}^n\right)$$

where

$$r = \frac{\Delta t}{\Delta x^2}. \tag{8.17}$$

During the corrector step, the provisional values are used, and the advection term is estimated as a backwards difference:

$$u_i^{n+1} = \frac{1}{2}\Big[u_i^n + u_i^{\overline{n+1}} - \frac{\Delta t}{\Delta x}\big(F_i^{\overline{n+1}} - F_{i-1}^{\overline{n+1}}\big)$$
$$+ r\big(u_{i+1}^{\overline{n+1}} - 2u_i^{\overline{n+1}} + u_{i+1}^{\overline{n+1}}\big)\Big]. \tag{8.18}$$

There is no simple stability requirement for the Mac-Cormack scheme, but Tannehill et al. (1997) recommend:

$$\Delta t \leq \frac{\Delta x^2}{|u|\,\Delta x + 2}. \tag{8.19}$$

Remember that these are dimensionless variables. These examples by no means exhaust the possibilities; literally scores of other schemes have been proposed to solve nonlinear equations of the Burgers type (cf. Fletcher, 1991; Tannehill et al., 1997; Hoffmann and Chiang, 2000).

Solution Scheme Accuracy

To evaluate the accuracy of these three schemes, we set the dimensionless space and time steps respectively to $\Delta x = 0.1$ and $\Delta t = 0.005$. Because nondimensional $Re_c = u\Delta x$, this choice results in a cell Reynolds number between 0 and 0.3, indicating viscosity dominates. All three schemes provide approximate solutions to the Burgers equation (figs. 8.4, 8.5, and 8.6). In the analytic solution, the initial form advects to the right ($x > 0$) and left ($x < 0$) and diffuses through time. The diffusive reduction in u means that the advection speed also slows, as can be seen in figure 8.4.

The schemes are of variable accuracy, as indicated by comparing specific calculated values with the equivalent analytic values at a given time step. Taking the differences and squaring and summing them results in the mean squared error of the finite difference (FD) solution. The formula is

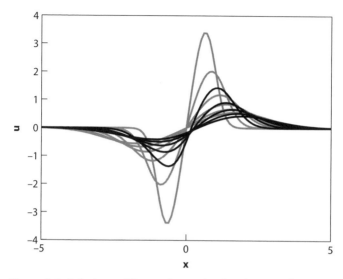

Figure 8.4. Solutions of Burgers' equation by FTCS explicit scheme for various times. Solid black lines are analytic solutions (equation 8.13); solid gray lines are numerical solutions with $\Delta x = 0.1$, $\Delta t = 0.005$. At these values, the cell Reynolds number and diffusion number (see text) meet their stability criteria, but the solution is still inaccurate.

$$\frac{1}{n}\sum_{i=1}^{n} e_i^2 \quad \text{where}$$

$$e_i = Y_i - \hat{Y}_i.$$

Y_i is the computed FD value at the ith space step, and $\hat{Y}_i$ is the analytic value. Figure 8.7 gives results for the examples presented above. As the solutions evolve in time, the errors decrease. The FTCS explicit solution scheme performs the worst, even though it is second-order accurate in space. The FTCS implicit scheme has a truncation error that is first order in time. The MacCormack scheme is second-order accurate in both time and space, explaining its better performance.

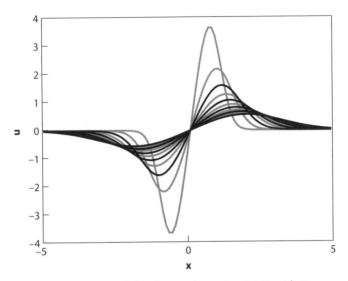

Figure 8.5. FTCS implicit solutions (equation 8.16) with $\Delta x = 0.1$, $\Delta t = 0.005$. Solid black lines are analytic solutions (equation 8.13); solid gray lines are numerical solutions.

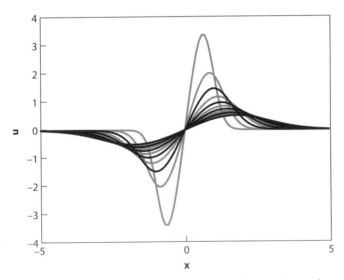

Figure 8.6. Solutions to equation 8.11 using the MacCormack predictor–corrector scheme, $\Delta x = 0.1$, $\Delta t = 0.005$. Solid black lines are analytic solutions (equation 8.13); solid gray lines are numerical solutions.

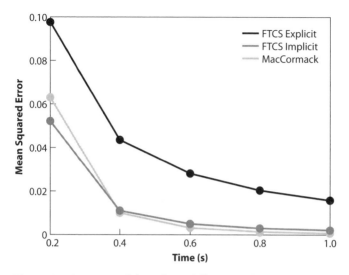

Figure 8.7. Accuracy of three finite difference schemes for solving Burgers' equation for the initial conditions given by equation 8.13 at $t = 0$. Mean squared error is the mean squared difference between the finite difference solution and analytic solution for all points in x. The FTCS explicit scheme performs the worst, as might be expected given its first-order accuracy in time.

Diffusive Momentum Transport in Turbulent Flows

In the derivation of Burgers' equation, we assumed that the flow was nonturbulent. Consequently, we assumed that momentum diffusion was a molecular process and the magnitude of momentum flux followed a first-order rate law (i.e., it was proportional to a velocity gradient). For Newtonian fluids experiencing viscous laminar flow this is correct, and the proportionality constant ν is a constant at standard temperature and pressure, depending only upon the type of fluid.

But there is a complication when modeling turbulent fluids. In addition to the Brownian motion of molecules, turbulence introduces another mechanism of momentum transfer—that due to the motion of *eddies*. Eddies are large bundles of fluid that maintain their coherency because they are rotating. They are generated in regions of increased shear, that is to say, where the fluid velocity gradients are high, after which they travel through the fluid while their spin rate slowly decays and they break apart into smaller eddies. Their mean motion during this travel occurs at a certain x-directed velocity u, and therefore they carry with them a packet of x-directed momentum equal to their mass, m times u. By analogy with molecular momentum transport, this can give rise to a net momentum flux and hence an increase in the apparent viscosity of the fluid. Consider some eddies of unit mass just upstream of, and some just downstream of, the face at x in figure 8.1. Their x-directed momentum in each case is given by their velocity, u. By turbulent motion, these eddies will be interchanging across the face at a certain rate. If their velocities are not the same, then in unit time there will have been an addition or loss of momentum from the control volume. The greater the difference in velocities across the face, the greater will be the addition or loss. Therefore, it seems sensible to assume that the flux of momentum across the face at x is proportional to the velocity gradient, or

$$q_{t_x} = -\mu_t \frac{\partial u}{\partial x}, \qquad (8.20)$$

where q_{t_x} is the diffusive momentum flux due to turbulence [kg m^{-1} s^{-2}], and μ_t is the proportionality constant [kg m^{-1} s^{-1}], and the negative sign ensures that momentum flows in the correct direction relative to the velocity gradient. μ_t is a viscosity [Pa s] due to the presence of turbulence in the flow. Its kinematic counterpart, μ_t/ρ, is called the *eddy viscosity* (eddy diffusivity, eddy-transfer coefficient, turbulent-transfer coefficient, or gradient-transfer coefficient), ν_t [m^2 s^{-1}]. Unlike the kinematic molecular viscosity, it is not a property of the fluid itself, but of the fluid *motion*, with values of up to 10^{11} times those of ν. Relating it to the mean motions of the turbulent flow is called the *turbulence closure problem*. A full discussion of the problem is beyond the scope of this book. Here we discuss the simplest possible realistic approximations to the eddy viscosity for geophysical flows. In geophysical flows, the horizontal length scale is many orders of magnitude larger than the vertical, and therefore ν_{th} is much greater than ν_{tv} so the two should be treated separately. In the horizontal, ν_{th} is often taken as a constant. In the vertical, if ν_{tv} is assumed to be constant, a parabolic velocity profile results that is clearly unrealistic for turbulent flows whose logarithmic profiles are well documented. From the heuristic description above, one might reasonably assume that ν_{tv} would scale with the size of the eddies and their turbulent intensity. Because the size is inhibited by the bed and water surface, and because the intensity of turbulence scales with u_*, ν_{tv} is often estimated as

$$\nu_{tv} = \kappa u_* z \left(1 - \frac{z}{H} \right), \tag{8.21}$$

where κ is von Karman's constant equal to 0.4 for clear turbulent water flows, u_* is the shear velocity, z is distance upwards from the bed, and H is total water depth. These concepts will be used in later chapters.

Adding Sources and Sinks of Momentum: The General Law of Motion

In the derivation of Burgers' equation, we ignored sources and sinks of momentum, writing the governing equation (equation 8.1) as if momentum were just another conservative scalar like mass. This allowed us to explore the behavior of the advection term. But an accurate description of geophysical flows requires including sources and sinks of momentum. These are usually expressed as forces on the mass of fluid in the control volume, such that equation 8.1 becomes

> Time rate of change of momentum =
> momentum rate in$_x$ − momentum
> rate out$_x$ + F$_x$. (8.22)

This is the general law of motion from chapter 1. The important forces for geophysical flows away from boundaries are (1) net pressure forces due to head and density differences and (2) gravity. Considering figure 8.1 and confining ourselves to only the x direction as before, we can conceive of the net pressure forces as the pressure on the face at location x minus the pressure on the face at $x + dx$, both multiplied by the area of the respective faces to convert a pressure to a force. The force of gravity on the mass in the control volume is simply mg_x. Thus equation 8.3 becomes

$$
\begin{aligned}
\frac{\partial \rho dx dy dz u}{\partial t} = {} & \rho dy dz uu - \left(\rho dy dz uu + \frac{\partial \rho dy dz uu}{\partial x} \right) dx \\
& + q_x dy dz - \left(q_x dy dz + \frac{\partial q_x dy dz}{\partial x} dx \right) \\
& + P_x dy dz - \left(P_x dy dz + \frac{\partial P_x dy dz}{\partial x} dx \right) \\
& + \rho dx dy dz g_x,
\end{aligned}
\tag{8.23}
$$

where P_x is the fluid pressure in the x direction. Dividing through by mass and extracting the terms expressing conservation of fluid mass as before yields

$$
\frac{\partial u}{\partial t} + u \frac{\partial u}{\partial x} - (v_m + v_t) \frac{\partial^2 u}{\partial x^2} = -\frac{\partial P_x}{\partial x} + g_x,
\tag{8.24}
$$

where both molecular and turbulent diffusion of momentum have been included. This is the general law of 1-D motion for an incompressible fluid. Remember that it is written for an infinitesimal control volume sitting in the middle of the fluid. To solve it, one must add two boundary conditions, one initial condition, and a domain of interest. Examples will be considered in subsequent chapters.

Summary

When the property being transported is momentum, the transport equation becomes nonlinear, thereby potentially generating higher-frequency wave modes and shock fronts. This poses additional problems for finite difference solution schemes. To be accurate, explicit finite difference schemes are limited to Courant numbers near one. Implicit schemes may be stable yet quite inaccurate at high Courant numbers.

Modeling Exercises

1. **Finite Difference Solution for a Lahar Flow**
   Equation 6.22 in chapter 6 gives the equation of motion of a lahar. Use the MacCormack scheme to solve for the temporally evolving thickness and velocity of a lahar flowing down a plane slope of 0.001 starting from an initial condition represented by a rectangle 5 m high by 200 m long. Assume the coefficient of drag is 1 and the kinematic viscosity is vanishingly small but nonzero.

2. **One-Dimensional Flow of an Alpine Glacier**
   Consider the 1-D glacier in figure 8.8. Let $h(x,t)$ be the height of the glacier above the datum, σ is the ice density (constant in space and time), $U = K\,h^n\,S^{n-1}$ where K is a dimensional proportionality constant, U is the cross-section average ice velocity [m s^{-1}], S is the slope of the ice surface, and q is the volumetric rate of addition of ice to the surface per

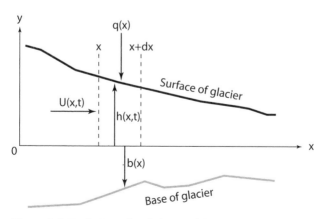

Figure 8.8. Definition sketch for modeling exercise 2.

unit length per unit time. The exponent n is a high power usually taken to be about 4. Derive a well-posed mathematical model describing $h(x,t)$, given $h(x,0) = F(x)$ and $q(x) = G(x)$.

3. **Solve the Viscous Burgers' Equation Over the Domain**

$-10 \leq x \leq 10$

with the boundary conditions $u(-10,t) = -2$ and $u(10,t) = 2$ and starting from the initial condition:

$$u = \frac{2 \sinh x}{\cosh x - e^{-0.1}}.$$

Use a numerical scheme of your choice. Compare your numerical results for different Peclet numbers with the analytic solution

$$u = \frac{2 \sinh x}{\cosh x - e^{-t}}.$$

Systems of One-Dimensional Nonlinear Partial Differential Equations

In earlier chapters, all of the problems involved one dependent variable as a function of one, two, or three independent variables. Here we introduce problems in which two dependent variables must be solved simultaneously to predict the evolution of a system. Examples include open channel flows in which the water velocity and depth are interdependent, lava flows in which the velocity depends upon a temperature-dependent lava rheology, and chemical systems in which two or more species are undergoing transport and reaction. Although systems of this sort are more complicated, their derivations require no new concepts; one need only ensure that the final well-posed problem contains N equations for the N dependent variables. On the other hand, finite difference solutions for systems of nonlinear PDEs are more complicated, and we devote proportionally more time to them here.

Translations

Example I: Gradually Varied Flow in an Open Channel

Predicting the velocity and water surface elevation of flows in a river or tidal channel is important for many reasons, such as flood forecasting and boating. For example, the United States National Weather Service is tasked with predicting flood routing along the nations's rivers and analyzing floodplains in the context of the National Flood Insurance Program. Their basic predictive tool is a 1-D model called FLDWAV. Here we derive a simplified version of FLDWAV. The resulting momentum equation is called St. Venant's equation after Adhemar Jean Claude Barre de Saint-Venant, the French mathematical physicist who first derived it. In simplifying, we assume without too much loss of accuracy that horizontal momentum diffusion (see chapter 8) can be ignored relative to advection.

Physical Picture

Consider a river reach of variable wetted cross-sectional area $A(x,t)$ [m²] as in figure 6.1. Flows routed through the channel are generally unsteady in time and nonuniform in the x direction. The velocity $V(x,t)$ and water depth $h(x,t)$ change with time and space and, consequently, so do the cross-sectional area, $A(x,t)$, and discharge, $Q(x,t)$. All variables are defined in table 9.1. By

Table 9.1. Variables for Chapter 9

$h(x,t)$ = cross-sectional average depth of flow [m]

$V(x,t)$ = cross-sectional average flow velocity in the x direction [m s⁻¹]

$Q(x,t)$ = flow discharge [m³ s⁻¹]

$A(x,t)$ = cross-sectional area [m²]

ρ = fluid density [kg m⁻³]

l = wetted perimeter [m]

f_{dw} = nondimensional friction coefficient $O(10^{-2})$

R = hydraulic radius [m]

P = average x-directed fluid pressure on the face at x [Pa]

q = rate of lateral inflow to channel per unit distance downstream [m³ s⁻¹ m⁻¹]

$T(x)$ = stream width [m]

τ_o = bed shear stress in the x direction [kg m⁻¹ s⁻²]

$$\begin{aligned} u_i^{n+1} = \frac{1}{2}\Bigg[u_i^n + \overline{u_i^{n+1}} - \frac{\Delta t}{\Delta x}\left(\overline{F_i^{n+1}} - \overline{F_{i-1}^{n+1}}\right) \\ + r\left(\overline{u_{i+1}^{n+1}} - 2\overline{u_i^{n+1}} + \overline{u_{i+1}^{n+1}}\right)\Bigg]. \end{aligned} \tag{8.18}$$

The 2-D, vertically integrated coastal ocean model derived above was used to investigate the response of the Lake Ontario circulation to wind forcing. The basin was discretized with a 4-km grid (fig. 10.7) and subjected to a persistent 4.7 m s^{-1} wind from the west.

The resulting surface currents are displayed in figure 10.8. Note that the resulting flows are quite similar to those shown in figure 10.2. There are strong easterly flows along the north and south margins of the lake and a general return flow along the deep central axis of the lake. The origin of this pattern is revealed by looking at the vertically integrated force balances on shallow versus deep water columns. In early stages of the simulation, wind shear stress drives water to the east everywhere, thereby creating a water-surface elevation gradient dipping to the west. After set up of the water surface gradient, the easterly flows are maintained on the shelves because the net eastward shear stress dominates over vertically integrated westward pressure gradient forces. In the deep axis of the lake, net wind shear stresses are reduced relative to pressure forces because they scale inversely with water depth. Relatively weak east to west flows develop, initiated by the water-surface elevation gradient and ultimately opposed by surface shear. Minor Coriolis accelerations created by north–south currents arising from topographic deflections of the largely east–west currents are important in the x-directed force balance in some regions of the lake. North–south force balances are largely geostrophic, with north–south water-surface gradients (set up by Ekman transport and topographically diverted flows) opposed by Coriolis accelerations arising from the east–west flows.

Summary

Modeling vertically averaged 2-D flows requires three equations: one conservation of mass equation to compute the water surface elevation and two conservation of momentum equations to compute the x- and y-directed velocities. The equations are a natural extension of Burgers'

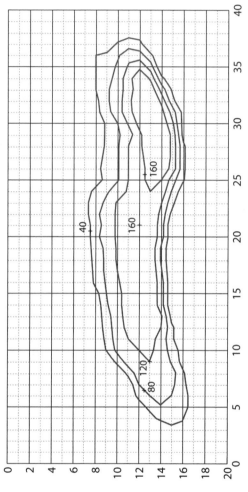

Figure 10.7. Gridded representation of Lake Ontario and bathymetry (depth in meters). Each grid is 4 km on a side. Axes represent computation node numbers.

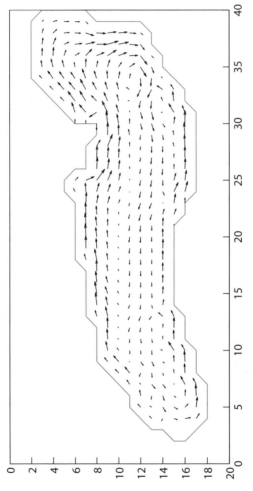

Figure 10.8. Velocity vectors showing vertically integrated steady-state Lake Ontario circulation in response to a uniform 4.7 cm s⁻¹ wind from the west. Vectors of one grid dimension are 0.1 m s⁻¹.

equation to which is added pressure, gravity, Coriolis, and bed friction sources and sinks of momentum.

Modeling Exercises

1. Role of Wind Velocity on Lake Circulation

Code the 2-D circulation model given above in your favorite computer language and conduct some numerical experiments on a circular basin where bathymetry monotonically deepens toward the basin center. Apply differing wind speeds and rationalize the resulting flows in terms of force balances.

Then generate your own discretized bathymetry for the lake of your choice and obtain average wind speeds and directions for the lake. Evaluate the resulting mean circulation and compare with known current patterns. Then obtain wind speed and direction data for a storm in this location, and simulate the surface current response to that event. Can you find observations to verify your analysis?

2. Two-Dimensional Flow in the Vertical

Using the derivation above, derive the equations describing 2-D flow in the vertical plane. Be mindful of how the gravity and pressure terms would change.

3. Geostrophic Flows along a Coast

Consider a coast in the Northern Hemisphere coincident with the y direction (y positive to the north) and x positive offshore (fig. 10.9). For simplicity, assume that the still water depth is everywhere equal and large enough such that bottom friction forces can be neglected. Further assume that a steady wind starts blowing to the south at time $t = 0$. We want to solve for the three dependent variables, $U(x, y, t)$, $V(x, y, t)$, and $h(x,y, t)$ as functions of the boundary and initial conditions. U and V are the vertically integrated flow velocities [m s^{-1}], and h is the dynamic water depth [m].

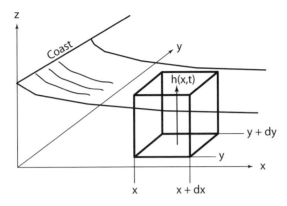

Figure 10.9. Definition sketch for modeling exercise
3. See text for variable definitions.

Derive the x-directed conservation of momentum
equation for the differential cell shown in figure
10.9. Assume turbulent diffusion of momentum is
negligible.

Closing Remarks

In this book we have attempted to teach geoscientists with little prior modeling experience how to translate common geological systems into mathematical models. Our goal was to instill a philosophy of science that exploits quantification as a valuable means of gaining insight. The book will have served well if readers incorporate thoughtful models in their day-to-day science and derive valuable physical understanding from them.

References

Aller, R. C. (1980). Diagenetic processes near the sediment-water interface of Long Island Sounds: I. Decomposition and nutrient geochemistry (S, N, P). *Advances in Geophysics* **22**:237–350.

Anderson, J. D., Jr. (1995). *Computational Fluid Dynamics: The Basics with Applications.* New York, McGraw-Hill.

Anderson, R. S., and S. P. Anderson (2010). *The Mechanics and Chemistry of Landscapes.* Cambridge, UK, Cambridge University Press.

Anonymous (1984). *CERC Shore Protection Manual.* Department of the Army, Waterways Experiment Station, Corps of Engineers, Coastal Engineering Research Center. Washington, DC: U.S. Government Printing Office.

Anonymous (1991). *Wind and Wave Climate Atlas, Volume III: The Great Lakes.* Prepared for Transportation Development Centre, Policy and Coordination Group, Transport Canada. Halifax, Canada, MacLaren Plansearch Limited.

Bear, J., and J. Buchlin (1991). *Modelling and Applications of Transport Phenomena in Porous Media.* Dordrecht, The Netherlands, Kluwer Academic.

Bellos, C., J. Soulis, et al. (1992). Experimental investigation of two-dimensional dam-break induced flows. *Journal of Hydraulic Research* **30**(1):47–63.

Berner, R. A. (1980). *Early Diagenesis: A Theoretical Approach.* Princeton, NJ, Princeton University Press.

Bird, R. B., W. E. Stewart, and E. N. Lightfoot (2006). *Transport Phenomena.* New York, John Wiley & Sons.

Boudreau, B. (1996). *Diagenetic Models and Their Interpretation.* Berlin, Springer-Verlag.

Chen, Y. (1979). Water and sediment routing in rivers. In: *Modeling of Rivers*. H. W. Shen, ed. New York, John Wiley & Sons, pp. 10.12–10.97.

Clark, M. (2009). *Transport Modeling for Environmental Engineers and Scientists*. Hoboken, NJ, John Wiley & Sons.

Crank, J. (1980). *The Mathematics of Diffusion*. Oxford, UK, Oxford University Press.

Csanady, G. (1982). *Circulation in the Coastal Ocean*. Berlin, Springer-Verlag.

Dietrich, W. E., D. G. Bellugi, et al. (2003). Geomorphic transport laws for predicting landscape form and dynamics. *Geophysical Monograph—American Geophysical Union* 135:103–132.

Dutton, J. A. (1987). Predicting the earth's future. *Acta Astronautica* 16:305–312.

Fletcher, C.A.J. (1991). *Computational Techniques for Fluid Dynamics*. Berlin, Springer-Verlag.

Fowler, A. C. (1997). *Mathematical Models in the Applied Sciences*. Cambridge, UK, Cambridge University Press.

Furbish, D. J. (1997). *Fluid Physics in Geology*. New York, Oxford University Press.

Gajdos, A., and S. Mandelkern (1998). Comparative study of numerical schemes used for one-dimensional transport modelling. Presented at: 2nd International PhD Symposium in Civil Engineering, Budapest, August 26–28, 1998. Available at: http://www.vbt.bme.hu/phdsymp/2ndphd/.

Garcia, M., and G. Parker (1991). Entrainment of bed sediment into suspension. *Journal of Hydraulic Engineering* 117(4):414–435.

Gisler, G. R., R. P. Weaver, et al. (2004). Two-and three-dimensional asteroid impact simulations. *Computing in Science & Engineering* 6(3):46–55.

Gomez, B. (1991). Bedload transport. *Earth-Science Reviews* 31(2):89–132.

Gregor, C. B. (1988). Prologue: Cyclic processes in geology, a historical sketch. In: *Chemical Cycles in the Evolution of the Earth*. C. B. Gregor, R. M. Garrels, F. T. MacKenzie, and J. B. Maynard, eds. New York, John Wiley & Sons, pp. 5–16.

Gyr, A., and K. Hoyer (2006). *Sediment Transport: A Geophysical Phenomenon*. Berlin, Springer-Verlag.

Haidvogel, D., and A. Beckmann (1999). *Numerical Ocean Circulation Modeling*. London, Imperial College Press.

Herschy, R. W. (2009). *Streamflow Measurement*. New York, Taylor & Francis.

Hoffmann, K. A., and S. T. Chiang (2000). *Computational Fluid Dynamics for Engineers*. Wichita, KS, Engineering Education System.

Hoffmann, K. A., J.-F. Dietiker, and A. Devahastin (2001). *Student Guide to CFD Volume I*. Wichita, KS, Engineering Education System.

Holland, H. D. (1978). *The Chemistry of the Atmosphere and Oceans*. New York, John Wiley & Sons.

Hornberger, G., and P. Wiberg (2005). *Numerical Methods in the Hydrological Sciences*. Special Publication Series 57. Washington, DC, American Geophysical Union.

Houghton, B., J. Latter, et al. (1987). Volcanic hazard assessment for Ruapehu composite volcano, Taupo volcanic zone, New Zealand. *Bulletin of Volcanology* 49(6):737–751.

Lassey, K. R., I. G. Enting, et al. (1996). The earth's radiocarbon budget. *Tellus* 48B:487–501.

Lee, N., V. Kourafalou, et al. (1985). Shelf circulation from Cape Canaveral to Cape Fear during winter. In: *Oceanography of the Southeastern U. S. Continental Shelf*. L. P. Atkinson, D. W. Menzel, and K. A. Bush, eds. Series on Coastal and Estuarine Regimes, Monograph 2. Washington, DC, American Geophysical Union, pp. 33–62.

Leonard, B. P. (1990). A stable and accurate convective modelling procedure based on quadratic upstream interpolation. *Computer Methods in Applied Mechanics and Engineering* (Special Edition on the 20th Anniversary): 59–98.

Le Roy, S. (2006). Numerical methods in the simulation of charge transport in solid dielectrics. *IEEE Transactions on Dielectrics and Electrical Insulation* 13(2):239–246.

Miller, R. (2007). *Numerical Modeling of Ocean Circulation*. Cambridge, UK, Cambridge University Press.

Murray, A. (2007). Reducing model complexity for explanation and prediction. *Geomorphology* 90(3–4):178–191.

Oberkampf, W., and T. Trucano (2002). Verification and validation in computational fluid dynamics. *Progress in Aerospace Sciences* 38(3):209–272.

Oreskes, N., K. Shrader-Frechette, et al. (1994). Verification, validation, and confirmation of numerical models in the earth sciences. *Science* 263(5147):641.

Paola, C. (2000). Quantitative models of sedimentary basin filling. *Sedimentology* 47(1):121–178.

Paola, C., P. L. Heller, et al. (1992). The large-scale variation in grain-size dynamics in alluvial basins, 1: Theory. *Basin Research* 4:73–90.

Parker, G. (1978). Self-formed straight rivers with equilibrium banks and mobile bed. Part 2. The gravel river. *Journal of Fluid Mechanics* 89(1):127–146.

Pelletier, J. D. (2008). *Quantitative Modeling of Earth Surface Processes.* Cambridge, UK, Cambridge University Press.

Pilkey, O. H., and L. Pilkey-Jarvis (2007). *Useless Arithmetic: Why Environmental Scientists Can't Predict the Future.* New York, Columbia University Press.

Preissmann, A. (1960). *Propagation des intumescences dans les canaux et rivierès.* Presented at: 1er Congrès d'association Française de calcul, Grenoble, France.

Richter, F. M., and K. K. Turekian (1993). Simple models for the geochemical response of the ocean to climatic and tectonic forcing. *Earth and Planetary Sciences Letters* 119:121–131.

Rothman, D. H., J. M. Hayes, et al. (2003). Dynamics of the Neoproterozoic carbon cycle. *Proceedings of the National Academy of Sciences of the United States of America* 100:8124–8129.

Simons, T. J., and W. M. Schertzer (1989). The circulation of Lake Ontario during the summer of 1982 and the winter of 1982/1983. *Environmental Canada, Scientific Series* 171:191.

Slingerland, R., J. Harbaugh, et al. (1994). *Simulating Clastic Sedimentary Basins.* Englewood Cliffs, NJ, Prentice Hall PTR.

Smolin, L. (2006). *The Trouble with Physics.* Boston, Houghton Mifflin.

Tannehill, J., D. Anderson, et al. (1997). *Computational Fluid Mechanics and Heat Transfer.* New York, Taylor & Francis.

Thuburn, J. (1996). Multidimensional flux-limited advection schemes. *Journal of Computational Physics* 123:74–83.

Turcotte, D., and G. Schubert (1982). *Geodynamics: Applications of Continuum Physics to Geological Problems.* New York, John Wiley & Sons.

Vernadsky, V. (1997; reprint). *The Biosphere.* New York, Copernicus.

Vignaux, M., and G. Weir (1990). A general model for Mt. Ruapehu lahars. *Bulletin of Volcanology* 52(5):381–390.

Walker, J.C.G. (1991). *Numerical Adventures with Geochemical Cycles.* New York, Oxford University Press.

Wigner, E. P. (1960). The unreasonable effectiveness of mathematics in the natural sciences. Richard Courant Lecture in Mathematical Sciences delivered at New York University, May 11, 1959. *Communications on Pure and Applied Mathematics* 13(1):1–14.

Willgoose, G. (2004). Mathematical modeling of whole landscape evolution. *Annual Review of Earth and Planetary Sciences* 33:443–459.

Yang, X.-S. (2008). *Mathematical Modelling for Earth Sciences.* Edinburgh, Scotland, Dunedin Academic Press Ltd.

Index

Accelerated Strategic Computing program, 6

accuracy, 5, 14; advection and, 122–23, 126; box modeling and, 60, 67, 71; finite difference and, 29, 33–38; hyperbolic systems and, 189; multidimensional diffusion problems and, 104; one-dimensional diffusion problems and, 78; one-dimensional partial differential equations (PDEs) and, 170, 175, 180–83; stability and, 35–37; transport and, 160–63

advection: accuracy and, 122–23, 126; backwards difference estimation and, 160; boundary conditions (BCs) and, 113–16, 121–22, 133–34, 138, 143, 149; checking units and, 114, 121, 133, 137–38, 142; Chézy equation and, 120; conservation of mass and, 111, 113, 119, 132, 135, 141; control volume and, 131, 135, 140–41; Courant-Friedrichs-Lewy parameter and, 123–24; Crank-Nicolson method and, 123, 126; defined, 111; dependent variables and, 112, 118, 121, 123, 132; derivatives and, 113–15; downslope gravity force and, 119–20; drag and, 120–21; finite difference and, 122–26; friction and, 119; hyperbolic equations and, 115; independent variables and, 112–18, 132; initial conditions (ICs) and, 114–16, 121–22, 124, 133–34, 138, 143, 147; interval definition and, 114–16, 121–22, 133–34, 138, 143; lahar channels and, 116–22; mass balance and, 114; momentum and, 111, 119, 130, 134, 136, 151–52, 155–60, 165; multistep (leapfrog) method and, 123; one-way wave equation and, 115; ordinary differential equations (ODEs) and, 115; parameters and, 146; partial differential equations (PDEs) and, 114–15, 122, 129–30, 170, 177, 179; Peclet number and, 95, 133–34, 144, 147, 151, 168; performing the balance and, 113–14, 119–21, 132–33, 136–37, 141–42; physical laws and, 113, 118–19, 132, 135–37; restrictive assumptions and, 113, 119, 132, 136, 140–41; river bed elevation and, 128–29; river pollutant and, 112–16; sedimentation of surface signal and, 128; Taylor series

advection (*continued*)
and, 113, 119, 124; translations and, 112–22; transport
and, 130–51
albedo, 53
Aller, Robert, 138–39
alternating direction implicit
(ADI) method, 106–8
Anderson, J. D., Jr., 27, 111
aquifers: multidimensional diffusion problems and, 96–99,
109; one-dimensional diffusion problems and, 75–80
Arakawa "C" grid, 197
Archimedes' principle, 21t
asteroids, 5–6
astrophysicists, 3

backward Euler method, 65–69
Bagnold, Reginald A., 136
Bear, J., 131
Beckmann, A., 187
Bellos, C., 181
Berner, Robert, 138–39
biosphere: climate models and,
2–3, 40, 48, 53–57, 90, 93,
187; cosmic rays and, 40;
photosynthesis and, 40–41,
48–50, 62; radiocarbon and,
39–48; sunspot cycle and, 45,
47. *See also* geoscience
bioturbation, 128, 140
Bird, R. B., 131
Boudreau, B., 89, 131
Boulding, Kenneth, 1
boundary conditions (BCs):
advection and, 113–16,
121–22, 124, 133–34, 138,
143, 149; box modeling and,
42–45, 50–52, 56–57, 62–65;
concept of, 1–2, 5, 9–13, 16;
Dirichlet, 12–13, 30, 78–79,
102, 183; elliptic, 101–8;
finite difference and, 23, 26,
30, 38; hyperbolic systems
and, 196–97, 201; initial
conditions (ICs) and, 26,
43–45, 50–52, 56–57, 63–65;

mixed, 13; multidimensional
diffusion problems and,
90, 95–96, 98–99, 101–2;
Neumann, 13, 34–35, 37, 78,
103; one-dimensional diffusion problems and, 74, 78–80,
83, 86–88; one-dimensional
nonlinear partial differential
equations and, 174, 177,
180–83; Robin, 13; transport
and, 133–34, 138, 143, 149,
155–57, 166, 168
boundary value problems (BVPs):
alternating direction implicit
(ADI) method and, 106–8;
case of variable coefficients
and, 107–8; explicit scheme
and, 102–3; finite difference
solutions for, 101–8; implicit
schemes and, 103–7
box modeling: automatic step-size
adjustment and, 70; backward
Euler method and, 65–69;
boundary conditions (BCs)
and, 42–45, 50–52, 56–57,
62–65; carbon cycle and,
39–40, 42, 48–52, 60–63;
checking units and, 42, 50,
56, 63; climate models and,
53–57; conservation of mass
and, 39–73; control volume
and, 39, 42, 54; defined, 39–
40; derivatives and, 47, 57–60,
64–66, 70; e-folding time
and, 44, 51; finite difference
solutions to, 57–70; forward
Euler method and, 57–59;
frequency and, 45–46; homogeneous reservoirs and, 39;
independent variables and, 57;
initial conditions (ICs) and,
43–45, 50–52, 56–57, 63–65,
71, 73; interval definition
and, 43–45, 50–52, 56–57,
63–65; matrices and, 66–70;
model enhancement and, 69;
ordinary differential equations
(ODEs) and, 39, 48, 51, 57,

70–71; performing the balance and, 42, 49–50, 55–56, 63; periodic forcing and, 45–47; phase lag and, 46–47; physical laws and, 40–42, 48, 53–55, 62; physical picture and, 40, 48, 53, 61–62; predictor–corrector methods and, 59–60; radiocarbon and, 39–48; residence time and, 45, 49–52, 61, 72; response time and, 40, 44–47, 50–52; restrictive assumptions and, 42, 49, 55, 62–63; Rothman ocean and, 61–69; sawtoothing and, 64–65, 68; steady state and, 40, 43, 56, 62–64, 68–73; stiff systems and, 60–61, 70, 73

Brownian motion, 76, 84, 136, 153, 163

bubbles in ice, 149

Buchlin, J., 131

Burgers, Johannes Martinus, 151–52

Burgers' equation: accuracy and, 160–63; analytic solution to, 157; finite difference and, 158–63; general law of motion and, 165–67; hyperbolic systems and, 203, 206; Newtonian fluids and, 152–56; solving over domain and, 168; staggered mesh and, 176–77; turbulent flows and, 163–65

burrowing organisms, 138–43

carbon cycle: box modeling and, 39–40, 42, 48–52, 60–63; decay and, 49, 59; dissolved inorganic carbon (DIC) and, 61–63, 68–69; dissolved organic carbon (DOC) and, 61–64, 69; photosynthesis and, 40–41, 48–50, 62; respiration and, 48; Rothman ocean and, 61–65

carbon dioxide, 149

Carolina Bight, 197

Cartesian coordinate system, 90, 97, 100, 189–90

causation, 3

centered in time and centered in space (CTCS) scheme, 30, 34–35

chain rule, 18

checking units, 10, 11t; advection and, 114, 121, 133, 137–38, 142; box modeling and, 42, 50, 56, 63; hyperbolic systems and, 196; multidimensional diffusion problems and, 95, 98, 101; nonlinear partial differential equations (PDEs) and, 171–74; one-dimensional diffusion problems and, 77–78, 82, 85–86; transport and, 133, 137–38, 142

Chézy, Anton, 120

Chiang, S. T., 23, 27, 29, 34, 160, 187

Chicxulub Crater, 5–6

circulation: Coriolis acceleration and, 21t, 192–97, 203, 206; hyperbolic systems and, 188–97, 202–3; Lake Ontario and, 188–90, 202–3; role of wind velocity and, 206

Circulation in the Coastal Ocean (Csanady), 187

Clark, M., 131, 135

climate models: albedo and, 53; biosphere and, 2–3, 40, 48, 53–57, 90, 93, 187; box modeling and, 53–57; heat exchange and, 56; hypothesis formulation and, 2–3; uniform temperature specification and, 56–57

coasts: circulation systems and, 188–97; evolution of coastline and, 80–83, 87–88; geostrophic flows and, 206–7

collisions, 3

Community Earth Surface Dynamics Modeling Initiative (CSDMS), 4–5 computational fluid dynamics (CFD), 187

Computational Infrastructure for Geodynamics, 5

conservation of energy, 21t, 100

conservation of mass, 10; advection and, 111, 113, 119, 132, 135, 141; box modeling and, 39–73; definition of, 20t; downslope gravity force and, 119; hyperbolic systems and, 191–92, 197, 203; lahars and, 119; multidimensional diffusion problems and, 90, 93–94, 97; one-dimensional diffusion problems and, 76, 82; partial differential equations (PDEs) and, 171, 175, 186; radiocarbon in biosphere and, 39–48; transport and, 154–55

conservation of momentum: advection and, 119; defined, 20t; hyperbolic systems and, 193, 203, 207; one-dimensional diffusion problems and, 85; partial differential equations (PDEs) and, 172, 176; transport and, 152, 154

control volume: advection and, 131, 135, 140–41; box modeling and, 39, 42, 54; concept of, 20; hyperbolic systems and, 193, 196; one-dimensional diffusion problems and, 77, 82, 85; partial differential equations (PDEs) and, 171–72; transport and, 152–53, 164–66

convection, 111, 126, 136, 144, 151, 158. See also advection

coordinate systems, 90, 97, 100, 139–40, 189, 199

Coriolis force: defined, 21t; hyperbolic systems and, 192–99, 203, 206

cosmic rays, 40

Courant-Friedrichs-Lewy parameter, 123, 201

Courant number, 123–24, 146, 159, 167, 177, 183

Crank, John, 75, 89

Crank-Nicolson scheme, 32, 35, 67, 86, 123, 126

Csanady, G., 187

cylindrical coordinate system, 139–40

dam-break problem, 180–83

Darcy, Henry, 97

Darcy's law, 21t, 75, 97–98, 109

Darcy-Weisbach equation, 97, 173

decay rates, 41

dependent variables: advection and, 112, 118, 121, 123, 132; box modeling and, 57–59; finite difference and, 23, 38; forward Euler method and, 57–59; hyperbolic systems and, 191, 194, 199, 206; multidimensional diffusion problems and, 90; one-dimensional diffusion problems and, 76, 78; partial differential equations and, 169–75, 178–79, 186; state variables and, 11–12; steps in model building and, 9–19; transport and, 152

derivatives: advection and, 113–15; box modeling and, 47, 57–60, 64–66, 70; finite difference and, 29–32, 37; hyperbolic systems and, 193; multidimensional diffusion problems and, 95, 103–4, 107; one-dimensional diffusion problems and, 74; partial, 12, 16–17, 70; steps in model building and, 12–19; time, 19

Diagenetic Models and Their Interpretation (Boudreau), 89, 131

Dietrich, W. E., 91

difference operators, 28–29

diffusion: advection and, 130–50; multidimensional,

89–110; one-dimensional, 74–88; Peclet number and, 95, 133–34, 144, 147, 151, 168; Reynold's number and, 21t, 137, 151, 159–60; transport and, 130–68 (see also transport); turbulent flows and, 163–65

diffusion number, 29, 37–38, 86–87, 159, 161

Dirichlet conditions, 12–13, 30, 78–79, 102, 183

discharge estimation, 147, 149

discretization: advection and, 122, 143; box modeling and, 58; finite difference and, 23, 26–27, 33–35; hyperbolic systems and, 203, 206; one-dimensional nonlinear partial differential equations and, 175, 183; transport and, 143

dissolved inorganic carbon (DIC), 61–63, 68–69

dissolved organic carbon (DOC), 61–64, 69

dissolved species, 111; advection and, 126, 128, 130; homogeneous aquifer and, 75–80; river pollution and, 112–16

dissolving sphere, 88

drag: advection and, 120–21; quadratic law of, 21t; transport of momentum and, 167

Dutton, John, 2–3

Early Diagenesis (Berner), 138

eddies, 163–65

e-folding time, 44, 51

elliptic boundary values, 101–8

empiricists, 2–3, 92

Enting, I. G., 43

environmental models, 3, 6, 131, 135, 197. See also geoscience

equations: advection, 113–15, 119–24; backward Euler method, 65–67; box modeling, 41–46, 48–51, 54–60, 63–67, 70; Burgers', 151–68;

chain rule, 18; Chézy, 120; Courant number, 123; downslope gravity force, 119; Fick's, 76, 140; finite difference, 26, 28–29, 31–32, 35; first-order, 74; forward Euler method, 57–59; functions and, 15; fundamental theorem of calculus and, 15–16; hyperbolic, 115, 192–96, 199–202; left-hand side (LHS) of, 30, 85, 113, 137, 154, 171, 194–95; matrix algebra, 24–25; multidimensional diffusion problems, 90, 92–95, 97–98, 100–8; multistep (leapfrog) method, 123; Navier–Stokes, 151; Neumann boundary condition, 13; nondimensionalization and, 13–14; one-dimensional diffusion problems, 76–77, 79, 81–82, 84–86; one-dimensional nonlinear partial differential equations (PDEs), 171–74, 176–80; one-way wave, 115; partial derivatives and, 16–17; performing the balance and, 42; periodic forcing, 45–46; predictor–corrector methods, 59–60; product rule, 18; radioactive decay, 41–42; right-hand side (RHS) of, 30, 46, 57, 85, 95, 98, 113, 132, 142, 154, 171, 173, 196; Robin boundary condition, 13; stepsize adjustment, 70; Taylor series, 19, 28–29; transport, 132–33, 136–38, 140, 142–43, 145–46, 152–60, 164–66; types of partial differential equations (PDEs) and, 18

error function, 36, 79

Euler solutions scheme, 151; backward, 65–67; box modeling and, 57–59, 64–69; forward, 57–59

extinction, 6

feedback, 2–3, 55, 185

Fick, Adolf, 76

Fick's law, 10, 21t, 75–76, 136, 140, 149

finite difference: accuracy and, 33–37, 160–63, 183; advection and, 122–26; alternating direction implicit (ADI) method and, 106–8; automatic step-size adjustment and, 70; backward Euler method and, 65–69; basics of, 26–33; boundary value problems (BVPs) and, 101–8; box modeling and, 57–70; Burgers' equation and, 158–63; centered in time and centered in space (CTCS) scheme and, 30, 34–35; consistency and, 33–37; Courant-Friedrichs-Lewy parameter and, 123–24; Crank-Nicolson method and, 32, 35, 123, 126; dam-break problem and, 180–83; derivatives and, 29–32, 37; discretization and, 23, 26–27, 33–35; explicit schemes and, 29–30; forward Euler method and, 57–59; implicit schemes and, 30–33, 38; Laasonen scheme and, 31; lahar flow and, 167; Lax equivalence theorem and, 34; matrices and, 23–26; model enhancement and, 69; modeling exercises for, 38; momentum and, 158–63, 166–67; multidimensional diffusion problems and, 101–8; multistep (leapfrog) method and, 123; one-dimensional diffusion problems and, 86; one-dimensional nonlinear partial differential equations (PDEs) and, 169, 175–83; ordinary differential equations (ODEs) and, 38; partial differential equations (PDEs) and,

23, 34, 101–8; physical picture and, 170–71; predictor–corrector methods and, 59–60; QUICKEST scheme and, 146–47; QUICK scheme and, 144–46; Rothman ocean and, 61–69; sawtoothing and, 64–65, 68; scheme assessment for, 33–37; stability and, 33–37; staggered mesh and, 175–77; steady state and, 36, 86; stiff systems and, 60–61, 70, 73; Taylor series and, 28–29, 34, 37–38; transport and, 143–47, 166–67; transport equation and, 143–47

first-order rate laws, 2, 21t, 74, 97

Fletcher, C.A.J., 23, 29, 34, 111, 160

flood prediction, 169–75

Florida, 7–8

flux: eddies and, 163–65; momentum and, 163–65; multidimensional diffusion problems and, 89–110; one-dimensional diffusion problems and, 74–88; transport and, 142, 152–56; turbulent flows and, 163–65

forward Euler method, 57–59

forward-in-time, centered-in-space (FTCS) scheme, 184; advection and, 122–23, 143; box modeling and, 58–59; finite difference and, 29, 33–38; hyperbolic systems and, 197, 200–2; multidimensional diffusion problems and, 102–6; nonlinear partial differential equations (PDEs) and, 175–77; one-dimensional diffusion problems and, 86; transport and, 158–63; two-dimensional vertically integrated geophysical flows and, 197–203

Fourier's law, 21t, 55, 75, 100

frequency, 14: advection and, 144; box modeling and, 45–46; fluid circulation and, 188; transport problems and, 166

friction: advection and, 119; Darcy-Weisbach, 173; hyperbolic systems and, 187, 192–93, 206; multidimensional diffusion problems and, 92–93, 97; partial differential equations (PDEs) and, 171–73, 179, 185

fudge factors, 3

fundamental theorem of calculus, 15–16

Gajdos, A., 144

gas eruption, 185

Gaussian elimination, 32

general law of motion, 165–67

Geodynamics (Turcotte and Schubert), 131

geoscience, 3–4, 22, 71, 89, 131, 209; aquifers and, 75–80, 96–99, 109; basic concepts for, 20–21; bubbles in ice and, 149; burrowing organisms and, 138–43; Chicxulub Crater and, 5–7; circulation systems and, 188–97; Coriolis acceleration and, 21t, 192–97, 203, 206; dam-break problem and, 180–83; discharge estimation and, 147, 149; dissolved species problems and, 75–80, 111–16, 126, 128, 130; gas eruption and, 185; geostrophic flows and, 206–7; gradually varied flow in open channel and, 169–75; gravity and, 119, 136, 166, 172, 206; Hurricane Ivan and, 7–8; lahars and, 116–22; landscape evolution and, 90–96; magma sill and, 150; pollutant transport and, 96–99, 109–16, 134–38; radioactive waste disposal and, 99–101; river bed elevation and, 128–29; sedimentary diagenesis and, 138–43; sedimentation of surface signal and, 128; suspended sediment in stream and, 134–38; tide propagation and, 185; transport problems and, 130–68 (*see also* transport); turbidity currents and, 185–86; volcanoes and, 185; wind velocity and, 206

Gisler, G. R., 5

glaciers, 167–68

Gomez, B., 91

gradient, 10, 19; advection and, 130, 138, 146; box modeling and, 55; finite difference and, 34; first-order rate laws and, 21t; hyperbolic systems and, 197, 199, 203; multidimensional diffusion problems and, 89, 93, 97, 107–8; nonlinear partial differential equations (PDEs) and, 177–78; one-dimensional diffusion problems and, 74, 77–78, 84, 87; transport and, 130, 138, 146, 152–55, 158, 163–64

gravity, 119, 136, 166, 172, 206

Gregor, Bryan, 46

groundwater elevation, 88

Gyr, A., 135

Haidvogel, D., 187

Hoffmann, K. A., 23, 27, 29, 34, 157, 160, 187

Holland, H. D., 45

Hooke's law, 21t

Hornberger, G., 89

Hoyer, K., 135

Hurricane Ivan, 7–8

hydraulics laws, 2

hyperbolic systems: advection and, 115; boundary conditions (BCs) and, 196–97, 201; Burgers' equation and, hyperbolic systems (*continued*) 203, 206; checking units and, 196; circulation and, 188–97, 202–3; computational fluid dynamics (CFD) and, 187; conservation of mass and, 191–92, 197, 203; control volume and, 193, 196; Coriolis force and, 192–99, 203, 206; dependent variables and, 191, 194, 199, 206; derivatives and, 193; dimensionality and, 187; friction and, 187, 192–93, 206; geostrophic flows and, 206–7; initial conditions (ICs) and, 196–97; interval definition and, 196–97; momentum and, 187–93, 196–98, 203, 206–7; partial differential equations (PDEs) and, 196–99; performing the balance and, 192–96; physical laws and, 191; physical picture and, 188–91; pressure and, 187–88, 192–97, 203, 206; restrictive assumptions and, 191–92; sinks and, 187, 206; steady state and, 196; translations and, 188–97; two-dimensional vertically integrated geophysical flows and, 197–203; unit outward vector and, 201–2; wind velocity and, 206

hypotheses, 2–3

independent variables: advection and, 112–18, 132; box modeling and, 57; functions and, 15; multidimensional diffusion problems and, 90; one-dimensional diffusion problems and, 76, 78; partial differential equations and, 169, 179, 186; state variables and, 11–12; steps in model building and, 9, 12–18; transport and, 132

initial conditions (ICs): advection and, 114–16, 121–22, 124, 133–34, 138, 143, 147; box modeling and, 43–45, 50–52, 56–57, 63–65, 71, 73; concept of, 2, 10, 12; defined, 10; finite difference and, 26, 31; hyperbolic systems and, 196–97, 206; multidimensional diffusion problems and, 90, 95–96, 98–99, 101–2, 105; one-dimensional diffusion problems and, 78–79, 83, 86; one-dimensional nonlinear partial differential equations and, 174, 181–83, 185; transport and, 133–34, 138, 143, 147, 151, 155–56, 163, 166–68

integration: advection and, 114; box modeling and, 41, 51, 65, 73; concept of, 12, 15–16, 18; hyperbolic systems and, 196–203, 206; multidimensional diffusion problems and, 90; one-dimensional diffusion problems and, 78; one-dimensional nonlinear partial differential equations and, 174, 186

interval definition: advection and, 114–16, 121–22, 133–34, 138, 143; box modeling and, 43–45, 50–52, 56–57, 63–65; concept of, 12, 14–16; hyperbolic systems and, 196–97; multidimensional diffusion problems and, 95–96, 98–99, 101; nonlinear partial differential equations (PDEs) and, 174–75; one-dimensional diffusion problems and, 78, 83, 86; transport and, 133–34, 138, 143, 155–56

Jacobians, 66–67, 69–70, 157, 159

kinetic energy, 6, 130

Laasonen fully implicit scheme, 31
Lagrange, Joseph-Louis, 19
lahars, 116–22, 167
Lake Ontario, 188–90, 202–3
landscape evolution, 90–96
Laplace equation, 89
Lassey, K. R., 43
law of mass action, 21t
law of radioactive decay, 21t, 40, 42, 89, 99
Lax equivalence theorem, 34
leapfrog method, 123
Lee, N., 197
Leonard, B. P., 144
Lightfoot, E. N., 131
linear systems, 25–26, 31, 47, 51–52
Long Island Sound, 139
Los Alamos National Laboratory, 5–6

MacCormack method, 159–61, 167
magma sill, 150
Mandelkern, S., 144
mathematics: abstract/reality issues and, 4; chain rule, 18; control volume concept, 20; functions, 14–15; fundamental theorem of calculus, 15–16; general solution and, 16; kinds of coefficients and, 18; matrices, 23–26; as miracle, 4; ordinary differential equations (ODEs), 15–17; partial derivatives, 16–17; partial differential equations (PDEs), 17–19; product rule, 18; Taylor series, 19; time derivatives, 19
Mathematics of Diffusion, The (Crank), 75, 89
MATLAB, 38, 68
matrices: box modeling and, 66–70; finite difference and, 31–32, 38; identity, 67; Jacobian, 66–67, 69–70; linear systems and, 25–26;

multidimensional diffusion problems and, 105–8; nonlinear partial differential equations (PDEs) and, 180
matrix algebra, 23–26, 38
Mexico, 5–6
Meyer–Peter Mueller bedload transport function, 91
Miller, R., 187
Modeling and Applications of Transport Phenomena in Porous Media (Bear and Buchlin), 131
models: abstraction/reality issues and, 4; as art, 2; balance and, 10; as big black boxes, 1; boundary conditions and, 1 (*see also* boundary conditions [BCs]); box modeling and, 39–73; causation and, 3; climate, 2–3, 40, 48, 53–57, 90, 93, 187; concepts for, 1–22; control volume and, 20, 39, 42, 77, 82, 85, 131, 135, 140–41, 152–53, 164–66, 171–72, 193, 196; dangers of, 3; dependent variables and, 9; derivatives and, 12–19; descriptors of, 2; determination of final state and, 2; deterministic, 1–2; differential equations and, 11–12; dynamical, 2–3, 14; empirical, 2–3, 92; enhancements to, 69; environmental, 3, 6, 131, 135, 197; feedback and, 2–3, 55, 185; finite difference and, 23–38; forward, 1; fudge factors and, 3; hypothesis formulation and, 2–3; independent variables and, 9; initial conditions (ICs) and, 2, 10 (*see also* initial conditions [ICs]); inverse, 1; logical terms and, 2; nondimensionalization and, 13–14; parameters and, 12, 14, 16, 67, 118, 123, 146; precision and, 3; predictions

models (*continued*)
from, 2–6, 8, 14, 44, 47, 59–60, 87, 99, 112, 117, 120–21, 129–30, 134–35, 151, 159, 169–70, 180–83, 188; proper posing of, 12; solving, 10; state variables and, 5, 11–12; steps in building, 8–21; stochastic, 1; structure-imitating, 1; symbol substitution and, 10; temporal/spatial changes and, 2; types of, 1–2; verification of, 10–11
momentum: advection and, 111, 119, 130, 134, 136, 151–52, 155–60, 165; Brownian motion and, 153, 163; Burgers' equation and, 151–68; conservation of, 20t, 85, 119, 152, 154, 172, 176, 193, 203, 207; convective term and, 151, 158; diffusion of, 83–86; drag coefficient and, 167; eddies and, 163–65; finite difference and, 158–63, 166–67; flux and, 75, 152–56, 163–65; general law of motion and, 165–67; glaciers and, 167–68; hyperbolic systems and, 187–93, 196–98, 203, 206–7; lahars and, 167; Newtonian fluids and, 152–56; Newton's second law and, 20t; one-dimensional nonlinear partial differential equations (PDEs) and, 170–72, 176, 179, 181; per unit mass, 152, 155, 164; pressure forces and, 151, 163–66; St. Venant's equation and, 151, 170, 175, 200; sinks and, 151, 153, 156, 165–66; suspended sediment and, 134–38; transport and, 130, 134, 136, 151–68; turbulent flows and, 163–65
Mount Ruapehu, 116–18
multidimensional diffusion problems: alternating direction implicit (ADI) method and, 106–8; aquifers and, 96–99; boundary conditions (BCs) and, 90, 95–96, 98–99, 101–2; boundary value problems (BVPs) and, 101–8; checking units and, 95, 98, 101; conservation of energy and, 100; conservation of mass and, 90, 93–94, 97; derivatives and, 95, 103–4, 107; elliptic boundary values and, 101–8; finite difference and, 101–8; flux and, 89–110; friction and, 92–93, 97; independent variables and, 90; initial conditions (ICs) and, 95–96, 98–99, 101; interval definition and, 95–96, 98–99, 101; landscape evolution and, 90–96; matrices and, 105–8; Meyer–Peter Mueller bedload function and, 91; partial differential equations (PDEs) and, 90, 100–1, 107; performing the balance and, 93–94, 97–98, 100–1; physical laws and, 90–93, 97, 100; physical picture and, 90; pollutant transport and, 96–99, 109–10; radioactive waste disposal and, 99–101; restrictive assumptions and, 93, 97, 100; steady state and, 89, 95, 99, 102–3; Taylor series and, 94, 103; translations and, 90–101; transport and, 96–99. *See also* transport
multistep method, 123
Murray, A., 3

National Flood Insurance Program, 170
National Science Foundation (NSF), 5
Navier–Stokes equations, 151
Neumann boundary conditions, 13, 78

Neumann stability analysis, 34–35, 37, 103
Newtonian fluids, 35, 152–56, 163
Newton's law of cooling, 150
Newton's law of motion, 20t
Newton's law of viscosity, 21t, 75t, 85, 153
Newton's universal law of gravitation, 21t
New Zealand, 116–18
Niagara River, 188
nitrogen, 40, 42
nondimensionalization, 13–14, 78–79
nonlinear systems, 40, 51
n-parameters, 16
Numerical Methods in the Hydrological Sciences (Hornberger and Wiberg), 89
Numerical Modeling of Ocean Circulation (Miller), 187
Numerical Ocean Circulation Modeling (Haidvogel and Beckmann), 187

Oberkampf, W., 11
oceanic P-cycle, 71–73
Ohm's law, 21t
one-dimensional diffusion problems: analytic solutions and, 79–80; boundary conditions (BCs) and, 74, 78–80, 83, 86–88; checking units and, 77–78, 82, 85–86; coastline evolution, 80–83, 87–88; conservation of mass and, 76, 82; conservation of momentum and, 85; conservative property and, 74, 86–87; control volume and, 77, 82, 85; derivatives and, 74; diffusion of momentum, 83–86; Dirichlet conditions and, 78–79; dissolved species in homogeneous aquifer, 75–80; Fick's law and, 75–76; finite difference and, 86; first-order rate law and, 74; flux and,

74–88; independent variables and, 76, 78; initial conditions (ICs) and, 78, 83, 86; interval definition and, 78, 83, 86; nondimensionalization and, 78–79; partial differential equations (PDEs) and, 76–79; performing the balance and, 77, 82, 85; physical laws and, 76–77, 81–85; physical picture and, 75–76, 80–81, 83; restrictive assumptions and, 77, 82–83, 85; steady state and, 86; Taylor series and, 77; translations and, 75–86. *See also* transport
open channel flow, 169–75
ordinary differential equations (ODEs): advection and, 115; box modeling and, 39, 48, 51, 57, 70–71; description of, 15–16; finite difference and, 34, 38; fundamental theorem of calculus and, 15–16; solution of, 15–16; steps in model building and, 12–16; systems of, 16
Oreskes, N., 3

Paola, Chris, 4, 92
parameters, 16; advection and, 118, 123; Courant–Friedrichs–Lewy, 123; defined, 14; differential equations and, 12; diffusion, 146; Euler methods and, 67
Parker, G., 92
partial derivatives, 12, 16–17, 70
partial differential equations (PDEs): advection and, 114–15, 122, 129–30; alternating direction implicit (ADI) method and, 106–8; boundary value problems (BVPs) and, 101–8; chain rule and, 18; checking units and, 171–74; conservation of mass and, 171, 175, 186; control

partial differential equations (PDEs) (*continued*)
volume and, 171–72; dam-break problem and, 180–83; dependent variables and, 169–75, 178–79, 186; description of, 17–18; elliptic, 18; finite difference and, 23, 34, 169, 175–83; friction and, 171–73, 179, 185; gradually varied flow in open channel and, 169–75; hyperbolic, 18, 196–99; independent variables and, 169, 179, 186; interval definition and, 174–75; kinds of coefficients and, 18; momentum and, 156, 170–72, 176, 179, 181; multidimensional diffusion problems and, 90, 100–1, 107; one-dimensional diffusion problems and, 76–79; one-dimensional nonlinear systems and, 169–86; one-way wave equation and, 115; parabolic, 18, 101–8; performing the balance and, 171; physical laws and, 171; product rule and, 18; restrictive assumptions and, 171; St. Venant's equation and, 170, 175; simultaneous solving and, 169; solution of, 18; staggered mesh and, 175–77; steady state and, 174; Taylor series and, 172; three basic types of linear, 18; tide propagation and, 185; translations and, 169–75; transport and, 130, 151, 156; turbidity currents and, 185–86; two-dimensional vertically integrated geophysical flows and, 197–203

Peclet number, 95, 133–34, 144, 147, 151, 168

Pensacola Bay, 7–8

performing the balance, 11t; advection and, 113–14, 119–21, 132–33, 136–37, 141–42; box modeling and, 42, 49–50, 55–56, 63; hyperbolic systems and, 192–96; multidimensional diffusion problems and, 93–94, 97–98, 100–1; nonlinear partial differential equations (PDEs) and, 171; one-dimensional diffusion problems and, 77, 82, 85; transport and, 132–33, 136–37, 141–42, 154–55

periodic forcing, 45–47

permeable seawalls, 88

phase errors, 144

phase lag, 46–47

phase speed, 201

photosynthesis, 40–41, 48–50, 62

physical laws, 4, 9; advection and, 113, 118–19, 132, 135–37; box modeling and, 40–42, 48, 53–55, 62; diffusion and, 140; hyperbolic systems and, 191; multidimensional diffusion problems and, 90–93, 97, 100; nonlinear partial differential equations (PDEs) and, 171; one-dimensional diffusion problems and, 76–77, 81–85; transport and, 140, 152–53

physical picture: advection and, 112, 118, 131, 135, 140; box modeling and, 40, 48, 53, 61–62; carbon cycle and, 48; hyperbolic systems and, 188–91; multidimensional diffusion problems and, 90, 96, 99–100; nonlinear partial differential equations (PDEs) and, 170–71; one-dimensional diffusion problems and, 75–76, 80–81, 83; radiocarbon content of biosphere and, 40; steps in model building and, 8–9, 11t; transport and, 131, 135, 140, 152

Pilkey, O. H., 3

Pilkey-Jarvis, L., 3
Poisson equation, 89
pollutants: confined aquifer and, 96–99, 109; dispersion of, 110; river advection and, 112–16; suspended sediment in stream and, 134–38
porosity, 97, 128, 141–42
predictor–corrector methods, 59–60
Preissmann, A., 177, 179
pressure, 5; hyperbolic systems and, 187–88, 192–97, 203, 206; momentum and, 151, 163–66; multidimensional diffusion problems and, 97; one-dimensional nonlinear partial differential equations (PDEs) and, 170–73, 177, 179
product rule, 18
Proterozoic Era, 61–62

quadratic drag law, 21t
QUICKEST scheme, 146–47
QUICK scheme, 144–46

radioactive decay, 40–44, 46
radioactive waste disposal, 99–101, 109
radiocarbon: box modeling and, 39–48; cosmic rays and, 40; decay rate of, 40–44, 46; periodic forcing and, 45–47; phase lag and, 46–47; production of, 40; sunspot cycle and, 45, 47
radiocarbon units (RCUs), 40–45
residence time, 45, 49–52, 61, 72
response time, 40, 44–47, 50–52
restrictive assumptions, 10, 11t; advection and, 113, 119, 132, 136, 140–41; box modeling and, 42, 49, 55, 62–63; hyperbolic systems and, 191–92; multidimensional diffusion problems and, 93, 97, 100; nonlinear partial differential equations (PDEs) and, 171;

one-dimensional diffusion problems and, 77, 82–83, 85; transport and, 132, 136, 140–41, 153
Reynold's number, 21t, 137, 151, 159–60
Richardson scheme, 34
Richter, F. M., 47
rivers: circulation systems and, 188–97; discharge estimation and, 147, 149; landscape evolution and, 90–96; suspended sediment and, 134–38
Robin boundary condition, 13
Rothman, D. H., 52
Rothman ocean: box modeling and, 61–69; Euler solutions scheme for, 64–65; performing the balance and, 63–65; physical laws and, 62; physical picture and, 61–62; Proterozoic Era and, 61–62; restrictive assumptions and, 62–63
Runge–Kutta scheme, 71

St. Lawrence River, 188
Saint-Venant, Jean Claude Barre, 170
St. Venant's equation, 151, 170, 175, 200
salt concentration, 88
sandy coastline evolution, 80–83
sawtoothing, 64–65, 68
Schubert, G., 131
Science Applications International Corporation, 5–6
sedimentary diagenesis, 138–43
sine waves, 121–22, 156
sinks: advection and, 122, 124, 130–34, 142–43; box modeling and, 42; hyperbolic systems and, 187, 206; multidimensional diffusion problems and, 100–1; nonlinear partial differential equations (PDEs) and, 186; one-dimensional diffusion

sinks (*continued*)
  problems and, 77–79; transport and, 130–34, 142–43, 151, 153, 156, 165–66
Slingerland, Rudy, 187
Smolin, L., 4
snow, 149
stability: accuracy and, 35–37; advection and, 123, 147; box modeling and, 61, 64–69; consistency and, 33; finite difference and, 33–37; hyperbolic systems and, 197, 201; momentum and, 158–61; multidimensional diffusion problems and, 103; one-dimensional nonlinear partial differential equations and, 175, 179; transport and, 147, 158–61
staggered mesh, 175–77
Star Trek (TV show), 46
steady state: advection and, 138; box modeling and, 40, 43, 56, 62–64, 68–73; finite difference and, 36, 86; hyperbolic systems and, 196; multidimensional diffusion problems and, 89, 95, 99, 102–3; nonlinear partial differential equations (PDEs) and, 174; one-dimensional diffusion problems and, 86
Stefan–Boltzmann law, 21t, 54
step-size adjustment, 70
Stewart, W. E., 131
stiff systems, 60–61, 70, 73
storm surges, 7–8
string theory, 4
sunspot cycle, 45, 47
suspended sediment in stream, 134–38

Tannehill, J., 152, 160, 187
Taylor, Brook, 19
Taylor series, 19; advection and, 113, 119, 124; finite difference and, 28–29, 34, 37–38; multidimensional diffusion problems and, 94, 103; nonlinear partial differential equations (PDEs) and, 172; one-dimensional diffusion problems and, 77; transport and, 154
tide propagation, 185
transport: advection and, 130–51; boundary conditions (BCs) and, 133–34, 138, 143, 155–56; bubbles in ice and, 149; Burgers' equation and, 151–68; checking units and, 133, 137–38, 142; conservation of mass and, 154–55; control volume and, 152–53, 164–66; convective term and, 136, 144, 151, 158; dependent variables and, 152; discharge estimation and, 147, 149; eddies and, 163–65; finite difference and, 143–47, 158–63, 166–67; flux and, 74, 75t, 136, 142, 152–56, 163–65; friction and, 151; general law of motion and, 165–67; glaciers and, 167–68; initial conditions (ICs) and, 133–34, 138, 143, 155–56; interval definition and, 133–34, 138, 143, 155–56; lahars and, 167; magma sill and, 150; Meyer–Peter Mueller bedload function and, 91; momentum and, 130, 134, 136, 151–68; multidimensional diffusion problems and, 89–110; Newtonian fluids and, 152–56; one-dimensional case for, 131–34; one-dimensional diffusion problems and, 74–88; partial differential equations (PDEs) and, 130, 151, 156; performing the balance and, 132–33, 136–37, 141–42,

154–55; physical laws and, 140, 152–53; physical picture and, 131, 135, 140, 152; plethora of processes described by, 130–31; pollutant, 96–99, 109–10, 112–16; QUICKEST scheme and, 146–47; QUICK scheme and, 144–46; restrictive assumptions and, 132, 136, 140–41, 153; Reynold's number and, 21t, 137, 151, 159–60; sedimentary diagenesis and, 138–43; sinks and, 151, 153, 156, 165–66; suspended sediment in stream and, 134–38; Taylor series and, 154; translations and, 131–43; turbulent flows and, 163–65

Transport Modeling for Environmental Engineers and Scientists (Clark), 131

Transport Phenomena (Bird, Stewart, and Lightfoot), 131

Trucano, T., 11

turbidity currents, 185–86

turbulence closure problem, 164–65

turbulent flows, 163–65

Turcotte, D., 131

Turekian, K. K., 47

United States National Weather Service, 169–70

universal law of gravitation, 21t

U.S. Army Corps of Engineers, 8, 188

U.S. Department of Energy, 6

verification, 10–11

Vernadsky, V., 40

Vignaux, M., 117–18, 120

volcanoes, 116–22, 185

von Karman's constant, 165

Walker, J.C.G., 70

waves: advection and, 115, 122, 126, 144; hyperbolic systems and, 201–2; one-dimensional diffusion problems and, 80–83, 88; one-dimensional nonlinear partial differential equations and, 180, 183; transport and, 144, 156, 166

Weir, G., 117–18, 120

Wellington–Auckland express, 117

Wiberg, P., 89

Wigner, Eugene, 4

Willgoose, G., 90

Wind River Range, 90–91

Yucatán Peninsula, 5–6